AF830987

Mine Waste Utilization

Mine Waste Utilization

Ram Chandar Karra
Gayana B. C.
Shubhananda Rao P.

CRC Press
Taylor & Francis Group
Boca Raton London New York

CRC Press is an imprint of the
Taylor & Francis Group, an **informa** business

First edition published 2022
by CRC Press
6000 Broken Sound Parkway NW, Suite 300, Boca Raton, FL 33487-2742

and by CRC Press
4 Park Square, Milton Park, Abingdon, Oxon, OX14 4RN

CRC Press is an imprint of Taylor & Francis Group, LLC

© 2022 Ram Chandar Karra, Gayana B.C and Shubhananda Rao P

Reasonable efforts have been made to publish reliable data and information, but the authors and publisher cannot assume responsibility for the validity of all materials or the consequences of their use. The authors and publishers have attempted to trace the copyright holders of all material reproduced in this publication and apologize to copyright holders if permission to publish in this form has not been obtained. If any copyright material has not been acknowledged please write and let us know so we may rectify in any future reprint.

Except as permitted under U.S. Copyright Law, no part of this book may be reprinted, reproduced, transmitted, or utilized in any form by any electronic, mechanical, or other means, now known or hereafter invented, including photocopying, microfilming, and recording, or in any information storage or retrieval system, without written permission from the publishers.

For permission to photocopy or use material electronically from this work, access www.copyright.com or contact the Copyright Clearance Center, Inc. (CCC), 222 Rosewood Drive, Danvers, MA 01923, 978-750-8400. For works that are not available on CCC please contact mpkbookspermissions@tandf.co.uk

Trademark notice: Product or corporate names may be trademarks or registered trademarks and are used only for identification and explanation without intent to infringe.

Library of Congress Cataloging-in-Publication Data
[Insert LoC Data here when available]

ISBN: 978-1-032-21455-9 (hbk)
ISBN: 978-1-032-21457-3 (pbk)
ISBN: 978-1-003-26849-9 (ebk)

DOI: 10.1201/9781003268499

Typeset in Times
by SPi Technologies India Pvt Ltd (Straive)

Dedicated To

Our Families

Contents

Contents

Foreword

Mining operations can generate enormous amounts of waste (i.e. waste rock and tailings) that must be managed and handled in an appropriate way to avoid any possible negative environmental impact, such as air, soil and water pollution. A lot of revenue is spent on waste rock excavation to mine ore or coal reserves. Waste occurs at several stages of the mining process in the life cycle of the mine, from exploratory drilling to mine closure. Thus, mining industry remains on top of the waste producers. However, not all waste generated from various mining activities are harmful to the environment.

The nature and character of mine waste depends on the type of the mineral extracted and the mining process. Typically, coal mines produce a lot of overburden waste rock whereas the metalliferous mining project generates a lot of mill tailings, the gangue material after the separation of useful ore concentrate. Of the several types of waste generated from these mines, waste rock and mill tailings stand out with the largest scope for reengineering to recycle.

It has become imperative to manage the mine-generated waste properly. In this context, a book on **'Mine Waste Utilization'** by Dr. Ram Chandar Karra with Dr. Gayana and Dr. Shubhananda Rao helps to understand the type of waste a mine produces and to plan the utilization of the waste; further, it serves as good reference book for both the industry personnel as well as those associated with the small-scale industries or factories in reusing mine waste. The book is also equally useful for PG and PhD scholars, as extensive literature is included and detailed experimental procedures are given to carry out various laboratory experiments to assess the suitability of mine waste for different purposes like bricks manufacturing, concrete pavements, vegetation, etc.

Dr. K. Umamaheshwar Rao
Director, NIT Rourkela
Former Director, NITK Surathkal
Professor
Department of Mining Engineering
Indian Institute of Technology Kharagpur

Preface

Extraction of minerals is essential for the growth of any country. But mining of minerals produces a large quantity of waste, which creates a lot of problems if not handled carefully. 'Waste' in general is defined as worthless, but mining waste can be used for different purposes. Though I am a hardcore mining engineer, my visits to some of the best mining universities in the world and to some of the old abandoned mining sites in different countries is an eye opening for me about 'mine-waste utilization'. So, I have started some basic research work in terms of mining aspects on utilization of mine waste. Though my core research is rock blasting and slope stability, situations made me to pay equal attention to work on mine waste utilization as some Masters and PhD scholars are keen to work with me. So, initially we have started to focus on coal mine waste. We have realized two major applications of coal mine waste, one is, utilization of sandstone in place of sand in concrete, though it will not give much advantage in terms of strength but its use reduces the burden of waste dump management for mining industry and on the other hand, the natural fine aggregates (sand) can be preserved. Second one is to replace cement with ash. In-fact, ash, which is a waste product from thermal power plants, was partially replaced in place of cement in concrete. In the past, ash disposal was a major issue, but today 'ash is cash', it is used in various applications. This research work is presented in Chapter 3. Another study was utilizing the sandstone for vegetation purpose using different additives like fly ash, bottom ash, sewage sludge and limestone powder. It was found that such coal mine waste can be effectively used for growing specific species, which will improve green cover in mining area; this research work is presented in Chapter 5. Laterite is another waste which is produced as part of small-scale quarrying operations, especially in the south western part of India. A large quantity of waste is being produced while making laterite bricks, and such waste creates a lot of environmental problems. Systematic study was carried out to use such waste also in concrete, and this research study is presented in Chapter 4. Another sector in mining which produces large quantity of waste is iron ore mining. Extraction of iron ore produces a large quantity of overburden waste and also waste is produced while processing the ore, called iron ore tailings (IOT). IOTs can be used in making bricks, but they will be denser due to high density of IOT. So, perlite, a volcanic rock which will be very lighter once heated at high temperature, is added to reduce the density of bricks. Perlite is also a thermal insulator. So in order to assess the efficiency of IOT–perlite bricks, a pilot-scale study was carried out by constructing model rooms. This research study is presented in Chapter 6. Another study was carried out on utilizing both iron ore waste and IOT in concrete pavements by replacing coarse aggregates and fine aggregates, respectively, this work is presented in chapter 7. This will result in reducing the burden of maintaining large mine waste dumps and tailing ponds by the iron ore industry and on the other hand mining of natural aggregates and river sand can be reduced as they are depleting resources.

Before each chapter, an extensive literature review is presented and general applications of mine waste are presented in Chapter 2. Brief introduction of mine waste, sources of mine waste, types of mine waste are presented in brief in Chapter 1.

This book gives complete information about different types of mine waste and its use for a beginner like an undergraduate student of mining or civil engineering. This book also gives an extensive literature related to utilization of mine waste, different experimental procedures, IS codes and approach to do systematic research for PG and PhD scholars. It is equally useful for academicians as a reference book with vast information, and it is also useful to the mining and construction industry to know the optimum mix percentage of different types of mine waste material for different applications.

Extensive research work of various other researchers is suitably quoted and acknowledged in this book in the form of references.

Dr. Ram Chandar Karra
Associate Professor of Mining Engineering
National Institute of Technology Karnataka, Surathkal, INDIA
Email: krc_karra@yahoo.com, krc@nitk.edu.in

Acknowledgements

Thanking everyone who helped in bringing out this book is not just a mere formality, but it's my responsibility. Writing a book involves the help from many individuals and organizations. I am very much thankful to the management of M/s The Singareni Collieries Company Limited, Telangana State, for allowing us to collect suitable samples from different locations. I am also thankful to the management of M/s MSPL Limited-Hospete, Karnataka, for allowing us to collect samples of iron ore waste and tailings. Thanks to the management of various small-scale laterite quarries in Dakshina Kannada District of Karnataka for giving samples.

I am thankful to my former MTech students Mr. Sainath Vanakuri, Mr. Manjunath B. and Mr. Vikas Chaitanya who have done extensive laboratory-scale studies to assess the suitability of different waste material for optimum results. I am thankful to Dr. Gayana and Dr. Shubhananda Rao who pursued their PhD under my guidance and also co-authors of this book for their extensive studies and systematic research. Their keen interest, quality publications and large research data have enthused me to initiate the process of writing this book.

I am very much thankful to Prof. K. Umamaheshwar Rao, Director of National Institute of Technology Karnataka, Surathkal, and a well-known academician in Mining Engineering for writing the Foreword for this book.

I am thankful to NITK administration and my colleagues & staff for their help at different stages. I thank Mr. D.C. Rai, Ms. Gayathri & other lab staff for helping in carrying laboratory experiments. I also thank Mr. Harivarsh Kotian – my M.Tech (by Research) student & Ms. B. Bharathi for typing and editing the chapters.

I am grateful to my parents and all my teachers because of whom today I am in my present position. The love, affection and patience shown by my wife and kids just cannot be expressed. I am blessed to have a highly qualified, loving, caring and dynamic wife Dr. Neelima who has scarified her illustrious scientific career from a prestigious research laboratory in the USA and returned to India because of me and who is always supporting, encouraging and motivating whenever I am down. Without her, there is neither professional life nor personal life for me. The love and affection shown by my kids Master Pranav Tej Reddy and Miss Nitika Shreya Reddy just cannot be expressed in words; they are very understanding and never disturb me even after my office hours, they too left their bright future and happy schooling in the USA and returned to India to join me.

I thank the publisher in bringing out this book in a very professional and attractive manner. I thank everyone who helped directly or indirectly in bringing out this book.

Dr. Ram Chandar Karra
Associate Professor of Mining Engineering
National Institute of Technology Karnataka, Surathkal, INDIA
Email: krc_karra@yahoo.com, krc@nitk.edu.in

Acknowledgements

Authors

Dr. Ram Chandar Karra, Associate Professor of Mining Engineering at National Institute of Technology Karnataka (NITK) Surathkal (Govt. of India), served as Head of the Department. His areas of research expertise are rock blasting, slope stability and mine waste management. Dr. Ram Chandar graduated in Mining Engineering from University College of Engineering, Kakatiya University, Telangana (formerly known as Kothagudem School of Mines), obtained MTech from IIT-BHU, PhD from NITK-Surathkal and post-doctoral fellowship from the University of Illinois at Chicago, USA. He also holds MBA degree in HR.

He was involved in eight research projects and 120 industry-sponsored consultancy projects related to rock blasting and slope stability in mining and infra projects. He has published around 100 research papers in National/International Journals/Conferences, produced four PhDs and six more scholars are working. He is the recipient of ***'SRG IT Award-2008–09'***, **'ISTE- SGSITS National Award**-2012', **'IE-India Young Engineer Award-2012', 'MEAI- Smt. Kiran Devi Singhal Memorial Award-2014', 'National Design Award- 2017 in Mining Engineering', 'Engineering Gold Medal-2018'** from MGMI. He is the Chief Editor of International Journal of Mining and Geosciences, consulting editor of **The Indian Mining & Engineering Journal and is on the editorial board of a couple of journals**. He is a Fellow of Institute of Engineers, life member of more than 10 professional societies and has visited the USA, Australia, Japan, UAE, Malaysia, Hong Kong, Nepal, etc.

Dr. Gayana B. C. has graduated in Civil Engineering, obtained her MTech degree in Transportation Engineering and received her PhD from the National Institute of Technology Karnataka, Surathkal, India, in the area of Mine Waste Utilization in Concrete Pavements. She has published around 20 Research papers in various National/International Conferences/Journals. She has travelled widely, such as to Japan, Singapore, Malaysia, to present her research findings. Dr. Gayana is a student member of International Society for Concrete Pavements (ISCP), American Society of Testing and Materials (ASTM) and American Concrete Institute (ACI), and affiliate e-member of American Society of Civil Engineers.

Dr. Shubhananda Rao P. is a Civil Engineer, obtained his MTech degree in Construction Technology & Management and PhD in the area of Mine Waste Utilization – in making thermal efficient bricks from mine waste National Institute of Technology Karnataka, Surathkal, India. He has published a couple of research papers and travelled widely to different countries like USA, Canada, Europe, Japan, etc. Dr. Rao is a fellow member of Association of Consulting Civil Engineers (INDIA), Professional Engineer of Engineering Council of India (ECI), member of the Institution of Engineers (India), member of the Indian Institution of Valuers, Registered Valuers Organization, recognized by Insolvency and Bankruptcy Board of India of asset class Land & Buildings and Registered Valuer of Insolvency and Bankruptcy Board of India (IBBI).

Vikas Chaitanya graduated in Civil Engineering and obtained MTech degree in Environmental Engineering from NITK, Surathkal. He has carried out his MTech dissertation work on the Assessment of Suitability of Coalmine Waste for Vegetation with Different Additives.

Manjunath. B is a graduate in Civil Engineering and has obtained MTech degree from NITK, Surathkal. He has carried out his MTech dissertation work on Utilization of Laterite for Partial Replacement of Fine Aggregates in Concrete.

Sainath Vanakuri is a graduate in Civil Engineering and has obtained M Tech degree from NITK, Surathkal. He has carried out his MTech dissertation work on Utilization of Coal Mine Waste for Partial Replacement of Fine Aggregates in Concrete.

List of Abbreviations

Al_2O_3	Aluminium oxide
Al	Aluminium
ASTM	American Society for Testing & Materials
ANOVA	Analysis of Variance
ANSYS	Analysis System
AAC	Autoclaved Aerated Concrete
BIS	Bureau of Indian Standards
CEC	Cation Exchange Capacity
Ca	Calcium
CaO	Calcium oxide
CO_2	Carbon dioxide
CO	Carbon monoxide
Cc	Coefficient of curvature
Cu	Copper
CuO	Copper oxide
CSEB	Compressed Stabilised Earth Blocks
CMC	Carboxy Methyl Cellulose
EC	Electrical Conductivity
EDX	Energy Dispersive X-ray
ECC	Engineered Cementitious Composite
EDX	Energy-dispersive X-ray spectroscopy
ELO	Epoxidized Linseed Oil
EPA	Expanded Perlite Aggregate
EP	Expanded Perlite
EPC	Expanded Perlite Concrete
FA	Fly Ash
gm	Gram
g/cc	Gram per cubic centimetre
GGBS	Ground Granulated Blast Furnace Slag
GHG	Green House Gases
Fe	Iron
Fe_2O_3	Iron oxide
IOW	Iron Ore Waste
IOT	Iron Ore Tailings
IS	Indian Standards
IRC	Indian Road Congress
IEA	International Energy Agency
KEL	Keltech Energies Limited
kg/m^3	Kilogram per cubic metre
PbO	Lead oxide

LCD	Liquid Crystal Display
LOI	Loss of Ignition
Mg	Magnesium
MgO	Magnesium oxide
MnO	Manganese oxide
D_{50}	Median diameter
CH_4	Methane
Mpa	Mega Pascal
MJ/kg	Mega Joule per Kilogram
MSW	Mine Spoil Waste
mm	Millimetre
Mt	Million tonnes
NMI	National Mineral Inventory
N/mm^2	Newton per square millimetre
NO_2	Nitrogen Dioxide
Ni	Nickel
O	Oxygen
OPC	Ordinary Portland cement
OB	Overburden
PPM	Parts per million
H_3PO_4	Phosphoric acid
K_2O	Potassium oxide
K	Potassium
PP	Polypropylene
ROM	Run Off Mine
STP	Sewage Treatment Plant
Si	Silicon
SEM	Scanning Electron Microscopy
SMB	Stabilised Mud Blocks
SBR	Styrene-Butadiene Rubber
Na_2O	Sodium oxide
Na	Sodium
Na_2SiO_3	Sodium silicate
SiO_2	Silicon oxide
SER	Society for Ecological Restoration
SEM	Scanning Electron Microscope
SOC	Soil Organic Carbon
SOM	Soil Organic Matter
k	Thermal conductivity
TGA	Thermo-gravimetric analysis
UNFC	United Nation Framework Classification for Resources
UCS	Uniaxial Compressive Strength
USDA	United States Department of Agriculture
UCS	Unconfined Compressive Strength
Cu	Uniformity coefficient
WHC	Water Holding Capacity

Wa	Water Absorption
W/mk	Watts per metre-kelvin
XRD	X-Ray Diffraction
XRF	X-Ray Fluorescence Spectroscopy
ZnO	Zinc Oxide

Mathematical Symbols

(+)	Addition
(-)	Subtraction
(/)	Division
(x)	Multiplication
>	Greater Than
<	Lesser Than
=	Equal to
( )	Round Bracket
[ ]	Square Bracket
Π	Pi
√	Square Root
%	Percentage
||	Modulus
°C	Degree Celsius
°F	Degree Fahrenheit
α	Alpha
μ	Micro

1 Introduction

Ram Chandar Karra

CONTENTS

1.1 INTRODUCTION

Mining of minerals is essential for the development of any country. More than 100 minerals are being produced around the world, including fuels, metallic, non-metallic, atomic and minor minerals. Among them, the highest quantity extracted is of coal and iron ore. In the last few decades, mining activity has been increasing tremendously to produce the required quantity of coal and non-coal deposits to meet the energy and infrastructure demands. Simultaneously, the problems faced by mining industry are also increasing with increased production. Mining is being blamed for causing land, water, and air pollution. Whenever a new mining lease is granted or any existing project is expanded, there is a protest from environmentalists or green activists blaming that mining activity is going to spoil the land. At the same time, it is a known fact that to meet the ever-increasing demand for coal and other metals, mining activity should be enhanced. So, optimum utilization of natural resources like the land usage needs to be optimized as the total land available is always constant.

There are basically two methods of mining, one is surface mining and the other is underground mining. Surface mining involves removal of complete waste rock to expose the ore body or coal. process of removing the waste bound to damage the natural ecosystem by producing various types of pollution. The waste produced is generally dumped outside the mine in the form of overburden dumps. These dumps occupy large amount of land, which loses its original use and generally gets degraded. Maintaining stability of dump is also a major issue for mining industry. Such waste material is produced at various stages of mining.

1.2 TYPES OF MINE WASTE

Mine waste can be classified as follows:

- Overburden: It includes the rock and soil particles removed while gaining access to the mineral deposit in surface mines. It will be deposited on the surface of mine itself, which occupies a large area.

DOI: 10.1201/9781003268499-1

- Waste rock: Materials that contain negligible amount of mineral concentration are termed as waste rock. It will not be economical to extract minerals from it based on the present technology. The waste rock is suitable for earthwork on site during mining operations and as aggregate for concrete works.
- Tailings: Tailings are mineral waste products and finely grounded particles generated during the processing of ore. It may also contain traces of processing chemicals and are deposited in the tailing ponds in the form of water-based slurry. Due to its fine particle size, tailings can be used only in selective operations. Based on the type of tailing ponds, the water can be drained so that the remaining waste can be dried.

1.3 PROBLEMS WITH MINE DUMPS

Overburden dumps change the natural land topography, affect the drainage system and prevent natural succession of plant growth (Bradshaw and Chadwick, 1980), resulting in acute problems of soil erosion and environmental pollution (Singh et al., 1994; Singh et al., 1996). The impact of mining waste can have lasting environmental and socio-economic consequences and can be extremely difficult and costly to address through remedial measures. Mining overburden has to be properly managed to ensure the long-term stability of disposal facilities and to prevent or minimize any water and soil pollution arising from acid or alkaline drainage and leaching of heavy metals (Zheng et al., 2016). Dump materials are generally loose, fine particles become highly prone to blowing by wind. These get spread over the surrounding fertile land, plants; disturb their natural quality and growth of fresh leaves (Arvind Kumar et al., 2011).

The unit amount of overburden that must be removed to gain access to a similar unit amount of coal or mineral material is known as stripping ratio (cu.m/t). In some of the projects, the stripping ratio has reached a double digit, and in future this will increase as deeper deposits to be extracted. In other words, to extract the same quantity of coal or ore more quantity of waste should be removed, which requires extra land to dump. The material is dumped in the form of decks of each 30 m height, in general, the decks ($3 \times 30 = 90$ m) can go up to 90 m, and in special cases up to 120 m is permitted. The dumps will have a natural slope based on angle of repose of the material dumped. United States Geological survey has estimated the land required to dump waste material based on dump height, slope angle, swell factor, density, etc. (Figure 1.1). For example, 500 Mt of dumping material requires around 200 hectares of base land, which can be dumped up to 150 m height with an angle of repose of 34°. This land requirement is in addition to the active mining area, such a vast area is lost forever and in addition, control of dust dispersion from the dumps, and maintaining the stability of the dumps is additional and unproductive work. So, if the waste produced from the mines is used for some other purpose, it is a great relief for not only to the company but also to the eco system. If the dump material is used for some other purpose, usage of large area can be avoided for dumps, waste storage and handling problems can be avoided for the mining company, the waste material is

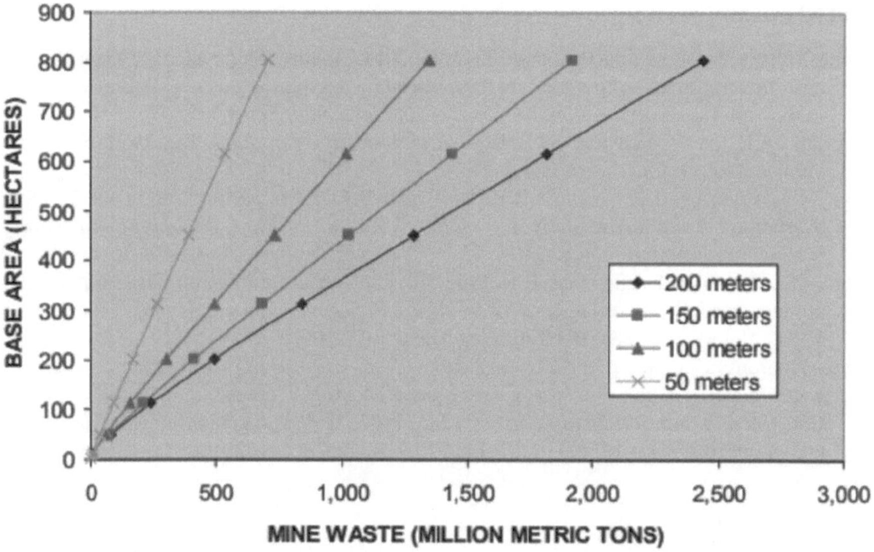

FIGURE 1.1 Mine waste base area required vs. waste material quantity for different height at 34° dump slope angle (https://pubs.usgs.gov/of/2003/of03-143/of03-143.pdf).

almost available at free of cost for the user. So, the different possible uses of mine waste are given in Section 1.4.

So, an effective waste handling and management system is essential. Attempt should be made as far as possible to reduce mine waste and tailings at generation site itself, available tailings to be used for some constructive purpose and the remaining to be disposed in an environmentally and effective way.

1.4 USES OF MINE WASTE

The possible utilization of mine wastes are as follows:

- As building material in construction industry
- For road construction
- For landfilling
- As an embankment material to mitigate traffic noise
- To stabilize pit walls or tunnels and to backfill stopes and galleries.
- For landscaping and soil stabilization during the closure of mine
- To neutralize acidic ground water generated in mine pits
- To build tailing dams
- As Fertilizers or supplements to enhance quality of land
- For back filling into surface mines
- For stowing into worked out underground mining areas

REFERENCES

Arvind Kumar, Rai, Biswajit, Paul and Singh, G. 2011. A study on physicochemical properties of overburden dump materials from selected coal mining areas of Jharia coalfields, Jharkhand, India. *International Journal of Environmental Sciences*, 1(6), 1350–1360.

Bradshaw, A.D. and Chadwick, M.J. 1980. *The restoration of land.* Blackwell Scientific Publications, Oxford, Great Britain.

Singh, R.S., Chaulya, S.K., Tewary, B.K. and Dhar, B.B. 1996. Restoration of a coal mine overburden dump a case study. *International Journal of Rock Mechanics and Mining Sciences and Geomechanics*, 33(8), 331.

Singh, R.S., Tewary, B.K. and Dhar, B.B. 1994. Effect of surface mining on plant biomass and productivity in a part of Dhanbad coalfield areas. *Second National Seminar on Minerals and Ecology*, Oxford & IBH Pub., New Delhi, 103–109.

Zheng, Juanrong, Zhu, Yalan and Zhao, Zhenbo. 2016. Utilization of limestone powder and water-reducing admixture in cemented paste backfill of coarse copper mine tailings. *Construction and Building Materials*, 124, 31–36. 10.1016/j.conbuildmat.2016.07.055. https://pubs.usgs.gov/of/2003/of03-143/of03-143.pdf Accessed 10th September 2021.

2 Utilization of Mine Waste

Ram Chandar Karra

CONTENTS

2.1 INTRODUCTION

Mining of minerals is essential to supply the raw materials for infrastructure development. There are two main methods of mining, namely surface mining and underground mining. In surface mining method, the overburden will be removed and dumped aside; the waste produced will be much higher than the coal or ore extracted based on stripping ratio (amount of waste to be removed in Cu.M to extract one tonne of coal or ore; this ratio could be more than 10:1 in some cases). The waste produced from various stages of mining should be disposed of in an environmental friendly manner. It should not cause any land pollution, water pollution, air pollution, etc. The waste dump yards should be stable. In order to avoid the environmental issues arising from the disposal of these wastes, it is better to use them for some other purposes.

DOI: 10.1201/9781003268499-2

Almost every coal as well as metal mining produces waste rock. In case of coal mines, the coal can be directly used as fuel, but in case of non-coal deposits, the ore can be processed using various methods and techniques like comminution, gravity separation froth floatation, etc., based on their physico-chemical properties (Ram Chandar et al. 2016a). Among the non-coal minerals, iron is the highest quantity produced around the world. The waste produced during the processing of iron ore is called iron ore tailings (IOTs), which is a waste.

2.2 UTILIZATION OF COAL MINE OVERBURDEN

Surface coal mining involves material that must be removed to gain access to the coal resources, including topsoil, overburden and waste rock (Zhengfu et al. 2010). In general, coal mine overburden contains shale and sandstone in layers. Shale is most abundant sedimentary rocks, representing nearly 80% of them. Mineralogically, shale is mainly made of montmorillonite (Kesavulu 2009). Montmorillonite is actually highly active clay (Punmia and Jain 2005). As using of clay in concrete is not possible, using of shale is also not possible. On the other hand sandstone contains more than 90% of the particles as sand, and it is possible to use sandstone in concrete (Kesavulu 2009).

A study on the coal mine overburden, in coal fields in India, where crushed over burden is used for stowing purpose, stated that there is 93% of the sand content present in the over burden and specific gravity of the over burden was found to be on an average 2.53. Compressibility characteristics were also found, and it was concluded that it is useful for stowing in underground mines, if the material washes through 0.15 mm sieve (Prashant et al. 2010). Another study from a different mine in India revealed that 98% of the particles are retained on 75 μm and higher sieve sizes (Rai et al. 2010). In this material the presence of nitrogen, potassium and phosphorus contents are very less and the soil is not useful for plantation (Yaseen et al. 2012). One more study carried out on the coal mine overburden reveals the sand content in the samples is more than 80% . The study suggested that the fresh mine spoil to attain the soil features of native forest soil through the process of reclamation shall take 28 years, provided the spoil habitat is not subjected to any other interferences like erosion, vegetational degradation, etc. (Maharana and Patel 2013).

2.3 UTILIZATION OF FLY ASH

Fly ash is finely divided residue resulting from the combustion of powdered coal and transported by the flue gases and collected by electro static precipitator. Fly ash is mostly used as Pozzolanic material all over the world (Shetty 2005). Manufacturers of reinforced concrete products commonly limit the quantity of fly ash to 25% or less by weight (Charles et al. 2005).

Unused fly ash is usually disposed into landfills contributing to soil, water and air pollution (Palomo et al. 1999; Duxson-Loya et al. 2007). As a product of the burning of coal, its type is determined by the type of coal. The anthracite and bituminous coals produce low calcium fly ash which possesses truly pozzolanic properties due to the high content of silica, while the lignite or sub-bituminous coals produce high calcium fly ash which is both a cementitious and pozzolanic material; it has lower silica and alumina but higher CaO content (Dhir et al. 1986).

There are two ways that fly ash can be used. One way is to mix certain amount of fly ash with cement clinker to produce Portland pozzolana cement and the second way is to use the fly ash as admixture at the time of making concrete at the site of work (Shetty 2005). American Society for Testing and Materials (ASTM) broadly classifies fly ash into two types. They are class F and class C. Class F is the fly ash normally produced by burning anthracite or bituminous coal. Class C is the fly ash produced by burning lignite or sub-bituminous coal (ASTM C618).

A high level of SO_3 in concrete can lead to volume instability and thus loss of durability, due to the formation of ettringite. Therefore, IS 3812-part-1(2003) allows for a maximum of 3% of total sulphur as SO_3 for fly ash to be used for concrete as binder. Finesses of the fly ash play an important role in the development of the mechanical strength (Fernandez-Jimenez and Palomo 2003).

2.4 UTILIZATION OF IRON ORE WASTE AND TAILINGS

In the form of steel, iron is the world's most commonly used metal, which is obtained by processing of iron ore. Iron makes about 95% of the total amount of metals used in a year (https://minerals.usgs.gov/minerals/pubs/commodity/iron_ore/mcs-2017-feore.pdf). It has applications in automobiles, machinery, marine purposes, structural engineering, etc. If we see the world reserves, out of 170,000 Mt of proved reserves, there exists only 83,000 Mt iron, and the remaining 87,000 Mt is going to be waste which is almost 51%, which indicates the steady increase in waste and reduction of iron content from iron ore. This increase in waste is due to extraction of high-grade ores in the past, and in future we need to extract the low-grade iron ore in order to meet the ever-increasing demand for steel. Considering the increase in rate of production, and decrease in iron content of ore in future, at least there would be 10% increase in the waste every year. In the processing of ore, different chemicals are added at different stages, which further increases the quantity of tailings slurry. So, constructing tailing dams every year to accommodate the increased tailings is a gigantic task, which requires a lot of space, construction materials and maintenance of the tailing dams. Instead, the tailings can be used for different purposes.

The main condition for mine waste or tailings utilization is that the material should satisfy all the geotechnical criteria and should be environmental friendly. A thorough characterization of mine waste is essential as it should not be a source of contamination. The value of utilization of mine waste can be enhanced on the basis of geotechnical properties and environmental constraints.

On the other hand, the natural resources like sand and coarse aggregates are becoming scarce and expensive due to the increased cost of transportation and depletion of river sand creating adverse environmental problems. Restrictions are made for the collection of river sand from river bed in some of the states in India, forcing concrete industry to look for an alternative for river sand. There are lot of restrictions coming up to extract aggregate from quarries which is essential in construction industry, as many of the quarries are closer to human habitats. Research is therefore required to investigate the use of cheaper, easily available and sustainable alternative materials to natural river sand like mine tailings and waste rock for replacement of aggregates.

Various researchers have studied the utilization of iron ore waste and tailings in various aspects including replacement of sand, aggregate or cement in concrete.

According to mineral council of Australia, key principles for effective mine tailing management are minimizing tailings production, adoption of a risk-based approach and increasing tailings reuse and considering relevant economic, environmental and social aspects (Joni et al. 2015). Mine tailing was used for ground improvement only after the beginning of the 20th century, and it proved to be an economic, environmental friendly and effective method of disposal.

2.4.1 Utilization of Iron Ore Tailings in Manufacturing of Bricks

Bricks are widely used in construction of buildings and other infrastructural facilities. Bricks are produced using clay and sand as raw materials and burnt in kiln at high temperature of 900 to 1,000°C. It is one of the primary building materials known to mankind. Over time, bricks have appeared, gained prominence and lost importance and then come to forefront with various styles of architecture. Burnt bricks were used in ancient India, Babylon, Egypt and Roman Civilizations (www.shodhganga.inflibnet.ac.in). They are still being used as filler materials for framework structures as well as to construct load bearing structures. The processes of making a brick traditionally are material tampering, moulding, drying, firing and sorting.

In general, the raw materials for production of bricks are clay and sand, which require considerable amount of energy in quarrying them. Clay quarrying adversely affects the landscape and produces some waste material. Bricks are burnt in intermittent kilns or continuous kilns. The kiln firing consumes significant amount of energy and releases huge quantity of green-house gases (GHG).

On the other hand, clay and sand are depleting resource and alternative materials and methods without firing to produce bricks will help in protecting the environment. Even though burnt bricks have better compressive strength, lower water absorption and less vulnerable to weathering, the non- burnt bricks are environmentally friendly eco bricks without consumption of depleting resources and having low embodied energy.

Clay is the main raw material for brick manufacture. Clay should exhibit some specific properties and characteristics for Civil Engineering applications. Clay must have sufficient moisture content and its air dried characteristics to maintain its shape after it is formed need to subject to optimum temperatures during firing and clay particles must fuse together.

The conventional method of making bricks causes serious environmental contamination, global warming, emission of GHG, smog, etc. Furthermore, energy as fuel and electricity showed a drastic consumption during the traditional manufacturing of bricks. As a result, lot of trees are cut in forests to use the wood as source of energy in firing stage of bricks production. Hence, recycling the wastes in the brick production appears to be viable solution for not only to environmental pollution but also economical option for design of green buildings.

Giri and Krishnaiah (Giri Babu and Krishnaiah 2018) reviewed on the utilization of waste materials for manufacturing of bricks to provide a potential and sustainable solution for the eco-friendly environment. Certain bricks are made without firing which is an advantage over other method of manufacturing of bricks in terms of low embodied energy material. It also offers a solution to the problem of waste disposal as well as eco-friendly environment in construction industry.

Use of IOTs to produce non-fired bricks not only can realize zero waste of IOT but also would offer a new raw material for building industry, which is more effective resource recovery alternative. In recent years, utilization of IOT has got much global attention especially in Civil Engineering Construction. This practice can help to reduce the emission of GHG: by avoiding the emission of virgin Engineering materials, providing cheaper alternative materials for building and constructions and also for natural resource conservation.

The IOTs contain high percentage of silica. The high silica content in the IOT is considered favourable in terms of the property and the raw material requirements for the production of ceramic tiles and bricks. They are unwanted material of economic interest from the gangue or wastes of IOTs and have other useful properties such as self-cementing characteristics which remove the necessity of adding cement when it is being used to fill mined out areas and for slimy nature, and it contains large quantities of sulphides which oxidize on contact with the air to form hard cement-like crust.

Hammond (1988) critically reviewed the use of mining and quarrying waste as building material. The availability, distribution and uses of waste from many mining countries of the world were discussed. The use of mining waste materials as concrete aggregates for the purpose of construction, production of brick and tiles, cement, pozzolana and pigments for paints were identified. It was stressed that by using waste materials, natural resources will be conserved, energy will be saved and environmental pollution will be reduced. Das et al. (2000) also described a new development in managing IOTs by converting them into value-added products such as ceramic floor and wall tiles for building applications. They reported that iron ore particles below 150 μm in size were discarded as waste tailings. They also tested constituents of the tailings from different locations and their mixture using standard techniques like XRD (Siemens D 500) with Ni filter and Cu (Kα) radiation. The high silica content in the IOTs is considered favourable in terms of the property and the raw material requirements for the production of ceramic tiles. The study concluded that IOTs up to 40% by weight can be considered for use as a part of raw materials for ceramic floor and wall tiles due to its high silica content. The ceramic tiles from the IOT materials were found to be superior in terms of scratch hardness and strength. The new tiles from the IOTs maintain most of the other essential properties as the conventional raw materials used for ceramic tiles. The application of IOTs in the ceramic tiles production was also found to be cost-effective in comparison with the usual traditional clay for ceramic tiles production.

Kumar et al. (2006) investigated the use of fly ash, blast furnace slag and IOTs in the production of floor and wall tiles. Different percentages of IOTs were tried with fly ash and blast furnace slag and proved that fly ash, blast furnace slag and IOTs in suitable combination in ceramic tiles will improve their qualities, including scratch hardness more than 6 on Mohr's hardness scale and flexural strength more than 25 MPa.

Ullas et al. (2010) found that water absorption of stabilized mud blocks (SMB) replaced by IOT is more but well within limits, i.e. 12–15%, because of increase in voids due to higher fine fraction. There is no linear expansion noticed when IOT is used. There is negligible fall of wet compressive strength of brick when IOT is used and it is only difference of 0.25 MPa when 100% sand is replaced by IOT.

Yongliang et al. (2010) determined the microstructure of fractured surface of the dried specimen using IOTs, cement, sand and gypsum without firing showing compact microstructure and sheet-like appearance of phyllosilicates and after firing to 1000°C for 2 hours; clear evidence of vitrification which was the typical grain and bond microstructure and crystalline phase were embedded in glassy matrix forming strong entirety, which promoted the strength of the brick. The compressive strength of finished brick was 15.9 MPa with ratio of tailings: cement: sand: gypsum = 78:10:10:2 with 15% forming water content. The strength of bricks decreased when IOT content was over 78%. Microstructure of the brick revealed the compact microstructure of sheet apperance of phychorilicates and the granule silica and hematite.

Ravikumar et al. (2012) studied replacement of IOT for cement in bricks and determined compressive strength; studies revealed that up to 15% of replacement of IOT will give higher compressive strength and replacement between 15% and 25% will give lower compressive strength. Studies conducted by Mangalpady (2012) revealed that the chemical composition of the tailings is directly dependent on the composition of the ore and mentioned about the suitability and reliability of IOT in the manufacture of pavement blocks. By using sand and IOT, a few reference mixes with cement, jelly, dust and baby jelly with different mix ratios were prepared. It was found that compressive strength of IOT-based mix was higher than the reference mix and the workability has improved up to 10% replacement of cement and IOT has pozzolanic property.

Carrasco et al. (2013) studied compression performance of walls of interlocking bricks made of iron ore by-products and cement. It deals with technical evaluation of the performance of walls constructed with interlocking bricks of iron ore by-products and cement under simple compressive loading. Three walls with dimensions of 150 cm width, 240 cm height and 15 cm thickness were built and tested. Testing was carried out according to the specifications of Brazilian standards. The first fissures arose with a stress of 0.56 MPa, corresponding to only 3.8% of the rupture stress of the brick alone. Horizontal displacement was negligible in all the walls and buckling was not observed. Rupture of the walls was through crushing; micro fissures appeared first and evolved into fissures and then transformed into cracks. After generalized occurrence of cracks, rupture occurred. This behaviour was similar to that of the bricks. Compressive load tests were also performed to determine the strength of the brick, of the prism (two overlaid bricks) and of the mortar. Results showed high compressive strength of 14.57 MPa for bricks, 9.82 MPa of the prisms and 25.2 MPa of the mortar. The walls showed good mechanical strength of 2.05 MPa, which represents 14% of the brick strength. Deformations were high, with axial deformation modulus of 420 MPa, which indicates a flexible behaviour of the wall. Although the wall is flexible, the fissuration stress is relatively high, indicating excellent performance of the wall. Another very positive aspect is that this stress is only 13.6% of the compressive strength of the wall and 1.9% of the brick, which indicates that there is a very large strength reserve.

Huang et al. (2013) developed green engineered cementitious composite (ECC), replacing cement by less reactive IOTs, reduced the matrix fracture toughness of ECC. Engineered cementitious composite will be of less compressive strength when replacement ratio is beyond 40% by IOT for cement. Mechanical properties and material greenness of ECC containing various proportions of IOT are investigated. IOT used in powder form with intention of enhancing the environmental sustainability of ECC. The ECC developed in this study with a cement content of 117.2–350.20 kg/m^3 has tensile ductility of 2.3%–3.3%, tensile strength of 5.1–6 MPa and compressive strength of 46–57 MPa after 28 days. The replacement of cement with IOT will result in 10%–32% reduction in energy consumption and 29%–63% reduction in carbon dioxide emission compared with typical ECC.

Francis et al. (2014) investigated the strength of the geo polymer bricks made from IOTs and sodium silicate (Na_2SiO_3), including UCS, durability and electrical resistivity. The strength of the geopolymer bricks made from IOTs with sodium silicate solution is influenced greatly by the curing temperature. The UCS increased as the curing temperature increased to a certain optimum point (80°C), and then the UCS decreased as the temperature increased further. The optimum base parameters for the production of the geopolymer bricks are sodium silicate solution content of 31%, initial setting time of 15 min and curing temperature of 80°C. The electrical resistivity of geopolymer bricks is lower than the commercial clay bricks due to the higher iron content associated with IOT. However, the electrical resistivity of the geopolymer bricks is still high enough to be used for building construction.

Jemish et al. (2014) conducted an experimental study on manufacturing of iron ore waste bricks made of iron ore fine wastes. Mixture was made of iron ore waste; sand and cement with four different proportions and attempts were made to find the suitability of the bricks. In this study, bricks with iron ore waste substituted for sand showed better compressive strength without much water absorption. The mixture made of cement, sand and iron ore waste with the ratio 30:30:40 and with 28 days curing period attained a compressive strength of 42.95 MPa and water absorption of 2.42%. This property met the requirement of IS 2180-1988 specification for heavy duty burnt clay building bricks in terms of compressive strength and water absorption.

Shreekant et al. (2016) carried out studies on utilization of iron ore waste in brick making for construction industry. He made an attempt to examine the possibility of making non-fired bricks from iron ore waste with some additives like cement and fly-ash. Each of the additives were mixed with iron ore waste (IOW) in different ratios and different sets of bricks were prepared. The prepared IOW bricks were cured for 7, 14, 21 and 28 days and their respective compressive strength and percentage of water absorption were determined. The results showed that IOW bricks prepared with 9% and above cement and with 28 days of curing are suitable for brick making and meet the IS specifications. It was also observed that the weight of the prepared bricks with 9% cement with 28 days of curing varies between 2.35 and 2.45 kg whereas the weight of compressed fire clay bricks varies from 2.80 to 2.89 kg. Results also showed that the cost of bricks prepared with cement ranging from 9 to 20% is comparable to that of commercially available compressed bricks.

Yisa et al. (2016) investigated the compressive strength of the laterite bricks with IOTs of weight 250 gm, 500 gm and 750 gm mixed with fixed quantity of soil, i.e. 2,000 gm from Zaria of Nigeria and arrived at a conclusion that compressive strength of laterite mix containing varying weight of IOT had higher strength value of 27 MPa when compared with only laterite which has compressive strength of 14 MPa for 7 days curing. Compressive strength of laterite bricks with IOT mix increased with the increase of IOT content compared to laterite.

Likith et al. (2017) studied manufacturing of building blocks by utilizing IOTs. The study presents that substitution of IOTs and quarry dust shows better compressive strength without much change in water absorption. It is also revealed that the mix with tailings has the highest compressive strength for 28 days curing. The use of these wastes instead of conventional materials not only preserves the natural precious resources but also solves the problems of disposal of waste.

Mendes et al. (2019) investigated on technical and environmental assessment of the incorporation of IOTs in construction clay bricks. The aim of this research is to study a new alternative for reusing high-silicon IOTs by applying it in the red clay ceramic industry. For this, a mixture design of experiments was developed, using three components, i.e. IOTs and two clayey materials. The two clayey materials are incorporated in ranges of 0%–40% and 30%–70% (by mass). Subsequently, ten mixtures obtained from the experimental design were prepared, and cylindrical specimens were formed by uniaxial pressing. After the firing at 850°C, 950°C and 1050°C, the properties of firing linear shrinkage, bulk density, apparent porosity, water absorption, compressive strength and microstructure behaviour of the specimens were assessed. Both the mixture design of experiments and desirable function enable the determination of an optimal composition which contains 29.1% (by mass) of tailings and meets the international standards. The brick has presented 20.94% of water absorption and compressive strength equal to 4.27 MPa, providing its potential to be used in sealing masonry.

2.5 UTILIZATION OF IRON ORE WASTE AND TAILINGS IN CONCRETE

Concrete is one of the most basic and critical components for any type of construction and plays an important role in building the nation's infrastructure. Concrete is a composite material which is composed of coarse and fine aggregates embedded in a matrix and bound together by a binder, which fills the space or voids between the aggregates (Mindess et al. 2003). Basically, concrete is a mixture of binder, water, aggregates and additives. The binder generally used in concrete production is OPC, which is mainly responsible for the mechanical strength. Utilization of a few industrial wastes as binder material could result in higher strength compared to OPC concrete strength. One of the major construction projects includes road construction which develops the country's infrastructure.

Research on replacement of sand with IOTs to prepare Ultra High Performance Concrete (UPHC) was conducted (Zhao et al. 2014). When the replacement level was not more than 40% for 90 days standard cured specimen, mechanical behaviour of tailings was comparable to that of control mix and the compressive strength decreased by less than 11% and flexural strength increased by 8% in comparison to control mix for specimen that were steam cured for 2 days.

Recent trends in autoclaved aerated concrete (AAC) has increased the requirement of waste utilization during the production of AAC. A large number of researches were carried out on the utilization of waste materials, such as fly ash, air-cooled slag, coal bottom ash, efflorescence sand, copper tailings and carbide slag, for the possibility of using them as AAC production (Zhao et al. 2014; Huang et al. 2013; Kurama et al. 2009; Mirza and Al-Noury 1986; Mostafa 2005). Further, AAC production was carried out using coal residues and IOT. Bulk density and compressive strength of prepared AAC were 609 kg/m^3 of 3.68 MPa, respectively. This AAC mainly composed of 20% CGC, 40% IOTs, 25% lime, 10% cement, 5% desulphurization gypsum and 0.06% aluminium powder (Wang et al. 2016). Preparation of lightweight tailings AAC block was recommended by Ma et al. (2015). AAC block has bulk density of 490 and 525 kg/m^3, compressive strength higher than 2.5 MPa with the composition consisting of cement, quicklime siliceous materials, gypsum and aluminium powder. Results for leaching test showed that AAC blocks with IOT were not a threat to environment.

Kumar (2014) studied on utilization of IOTs as replacement to fine aggregates in cement concrete pavements. In this study, IOT used as partial replacement to fine aggregates at levels of 10%, 20%, 30%, 40%, 50% and the basic material properties, strength parameters are studied. It is found that as the IOT percentage increases in the mix, workability is reduced. The cube specimens were tested in Compression Testing Machine after specified curing period for different percent of IOT replacement Mix1(10%IOT), Mix2(20%IOT), Mix3(30%IOT), Mix4(40% IOT) and Mix5(50%IOT) and for normal concrete mix. At 40% replacement level for 28 days, compressive strength is more than the reference mix (Normal concrete mix) and other replacement percentage mixes. Flexural strength is observed maximum for reference mix. Quality of concrete mixes is found good from Ultrasound Pulse Velocity test. Flexural fatigue analysis is carried out on mix with 40%. IOT replacement at stress ratios 0.65, 0.70 and 0.75 compared with IRC model for number of repetitions using log normal distribution, up to 0.7 stress ratio, it showed more number of repetitions than IRC and at higher stress ratio mix with IOT achieved failure earlier.

Prahallada and Shanmuka (2014) studied on stabilized IOT blocks showing an increasing trend in the erosion resistance with increase in curing period. Stabilized IOT blocks showed decrease in liquid absorption with increased stabilizer percentage and curing period, i.e. 1.35% on 21 days curing of 7% cement stabilization. Maximum dry compressive strength of IOT blocks with 7% cement stabilization on 21 days curing is 8.5 MPa, and the ratio of wet to dry compressive strength lies between 0.50 and 0.73.

Prem Kumar et al. (2014) experimented by replacing 0%, 10%, 20%, 30%, 40%, 60%, 80% and 100% of sand by IOT. There is an increase in the compressive strength due to sand replacement by IOTs. The maximum increase in compressive strength for about 40% of sand replacement, and there is no reduction in flexural strength of reinforced concrete beams.

Ugama and Ejeh (2014) studied the suitability of IOT as fine aggregate for replacement of sand in masonry mortar and found compressive, tensile and flexural strength of 36.95 MPa, 1.76 MPa and 5.73 MPa, respectively, for optimum level of 20% of IOT replacement.

Abdulrahman (2015) produced sand concrete blocks with mix ratio of 1:6 (one part of cement to six parts of sand) where sand portion was replaced by IOT with different percentages like 10%, 20% and 30% at 28 days curing; compressive strength approached the recommended strength for 230 mm blocks and proved that there is a way for waste disposal (IOT) and development of eco-friendly sand Crete blocks.

Kshitija et al. (2015) studied the use of IOTs as a construction material. In the study, IOTs are used as a partial replacement of the fine aggregates. In this study, 15%–20% replacement of IOTs is recommended which can save up to 20% of fine aggregates, thereby reducing the cost of production and also pollution of environment by using IOT and marching towards sustainable construction. Concrete with IOTs is a sustainable solution as it reduces sand by as much as 15% by IOTs in concrete.

Nagaraj et al. (2016) have conducted exploratory study on compressed stabilized earth blocks (CSEB) utilizing various proportions of mine spoil waste (MSW) (accumulated at up stream of mining area at Sandur region, Karnataka), quarry dust and stabilizers (cement and lime). Mine spoil waste was utilized in three possibilities 30%, 40% and 50% with cement and lime as stabilizer in two combinations like 6% cement + 2% lime and 8% cement + 2% lime in CSEB blocks. It was observed that wet compressive strength for any combination of admixture is more for blocks prepared with 40% MSW, which indicates these blocks can be effectively used as eco-friendly bricks in construction industry.

2.6 UTILIZATION OF MINE WASTE AS BACKFILL

Waste rock and mine tailings are extensively used in the backfilling of underground mined out areas, this could help in decreasing the amount of land used for mining activity and reduce the environmental impact (Lu and Cai 2012). The process of back filling involves deposition of tailings and waste rock with required amount of water and binders into the empty stopes. After few months of curing, the backfill will be able to provide adequate support to the adjacent stopes allowing further mining operation. In most cases, about 40% to 60% of the tailings produced in a mine can be reused as backfill, thereby reducing the environmental footprint left behind by the surface impoundment of tailings (Barry and James 2016).

Three most widely applied backfills are hydraulic fill, rock fill and CPB. Hydraulic fills are produced using de-slimed tailings and use binders to acquire mechanical strength after the curing period. Usually waste rock is used for dry back filling and tailings for hydraulic (Lu and Cai 2012). However, the paste – backfill had to be optimized for each site as the properties of tailings and water had significant effect on the strength acquisition of backfill. However, the back filling is an unproductive work which incurs additional cost. Back filling is very useful in case of weak host rock where cut and fill method of stoping is adopted; in case of sublevel stoping, the host rock is generally very hard and does not need any additional support; in that case, the dried tailings or crushed waste rock can be dumped in stopes.

Wei et al. (2018) used waste rock-tailing paste to control subsidence in earthquake-prone areas and secondary disasters, like as down-hole debris flow and groundwater contamination at Tong Cheng mine, China. Among the three backfill methods adopted (All non-cemented backfill, All-cemented backfill and Cemented–non-cemented joint backfill), the cemented–non-cemented combination was selected after their comparison. A backfilling strategy of vertical stratification and horizontal partition was developed for the subsidence area. The backfilling paste had a mass concentration of 81%–83%, 25%–30% of waste rock and 2%–5% of cement. The paste showed satisfactory anti-disintegration property and attained a compressive strength of about 1.5 to 3 MPa in 28 days. When the coefficient of permeability of the backfill ranged between 10^{-4} cm/s and 10^{-3} cm/s, a moderate level of permeability was achieved and this was suitable for the control of subsidence.

2.7 USE OF IOT IN CEMENTED PASTE BACKFILL

The Cemented Paste Backfill (CPB) technique is an efficient method to dispose the mine tailings in a cheaper way by backfilling of open pits and underground stopes in mines. CPB establishes a fair amount of ground support to the structures in the mine sites, thus facilitating the full excavation and complete extraction of ore deposits. It also minimizes the surface impoundment of tailings, thereby reducing the rehabilitation/reclamation cost involved during closure.

The use of CPB for disposing the acid-producing mine tailings instead of traditional disposal methods are opening new scopes in mineral waste management studies. The key benefits of backfilling using CPB are as follows: (a) filling of voids created by mining activities and thus providing regional and local stability to the ore deposit while facilitating an economical and safe tailing disposal system; (b) around 60% decrease in the surface impoundment in the sulphidic waste, thus minimizing the environmental pollution; (c) increase in the amount of ore that can be extracted by providing more support to ground by stabilizing the mined out areas; (d) reduced stope cycle time as there is not much free water to be drained out during the curing period. Hence, CPB can be used as a building material to build a wall to mine out the deposits (Amjad and Ernest 2013).

CPB is being used for surface disposal also. The key benefits include the following: (a) better hydro-geotechnical properties of tailings; (b) in underwater storages; (c) reduced particle segregation due to the homogeneity of the paste; (d) improved durability and strength of resultant support; (e) due to the addition of alkaline binders, the CPB matrix possess the potential for acid neutralization as well as contaminant stabilization (Amjad and Ernest 2013).

In 2007, environmental desulphurization (ED) was combined with CPB technology in the Doyon gold mine of Quebec, Canada. CPB became popular around the world over the last few years due to the environmental as well as operational benefits of mine waste backfilling technology. The Doyon tailings, even though contains only small amount of sulphide, are categorized as acid-generating, while the desulphurized version is categorized as non-acid generating. The CPB made with desulphurized tailings alone or part of the latter showed improved mechanical strength.

The study proved that in the case of Doyon mine, it is able to produce non-acid generating sulphide concentrates using ED technique. The sulphide concentrate could then be used in CPB. ED technique also assists the strength-gaining process in some cases (Benzaazoua et al. 2008).

In 2013, an experimental study was conducted using sulphide mill tailings to find the effects of stress conditions as well as curing on geotechnical, chemical and mechanical properties of CPB. The samples for the study were collected from a mine in Quebec, Canada. Consolidated, drained and undrained CPB samples were made with 3%, 4.5% and 7% by weight of binder content using different moulds. Samples were cured for 7, 14 and 28 days. CUAPS (Curing Under Applied Pressure System) was used to simulate the consolidation process. The backfills thus consolidated to have a compressive strength of about 1.5 to 2.8 times higher than that of the conventionally drained backfills. The pressure application during the initial stages of curing simulated the site condition which consequently increased the CPB strength. For a given binder content, the specific gravity, void ratio and degree of saturation of the CPB samples decreased with increase in curing time. The consolidation process was found to be increasing the stiffness of the CPB paste. This can be attributed to the rearrangement of the particles and resultant reduction in void ratio. Hence, the conventional plastic moulds can be replaced with CUAPS consolidated backfill samples to get a more practical CPB mix design. This method helps in cost optimization in backfill technology (Erol et al. 2014).

Expansion properties of CPB have to be considered while using sulphide tailings for the development of CPB. The effects of different sulphur content, curing time, solid concentration and cement dosages were analysed to find their effects on free expansion properties of CPB. The test results indicate that the free expansion ratio maintains a linear relationship with the sulphur content. The results also showed that the expansion behaviour mainly started only after 28 days of curing (Yin et al. 2018).

Promising results were obtained from another experimental study on the combined effect of limestone powder (LP) and WRA (Water-reducing admixtures) on the properties of CPB made from copper mine tailings. The tailings were classified as coarse sized as the particle size analysis showed that 18.73% weight of the sample was finer than 20 μm in size. The tailings had a sulphide content of only 5.14% by weight. The major miner contents were pyrite, muscovite, clinochlore, gypsum, quartz, calcite, dolomite, siderite and hematite in which the major sulphide mineral was pyrite. A complex binder which composed of 80% by weight was GBFS (Granulated blast furnace slag) and 20% OPC was used in the study. The limestone powder was made by grinding the by-products from quarry crushers. Various workability tests were conducted, and the unconfined compressive strength (UCS) of various specimens were tested. Also XRD (X-ray diffraction) and MIR (Mercury Intrusion Porosimeter) studies were conducted on the samples. It was found out that increasing the LP dosage to 10% by weight increased the workability of CPB paste. But beyond this dosage, the slump of the mixture found to decrease. This combination of LP and WRA improved the mechanical performance and packing density of CPB made from copper mine tailings. The long-term stability problems (i.e., loss of strength) were found to be subsided by the combined effect of LP and WRA as the UCS increased (Zheng et al. 2016).

2.8 USE OF IOT IN SOIL STABILIZATION

In recent years, mine tailings are also being used for the stabilization of weak soils such as black cotton soil (BCS), stabilization of laterite soil and also for the stabilization of base during road construction.

Tropical black clay or BCS is a highly expansive soil, which offers greater difficulty in construction over it. From long back, cement has been used for the stabilization of BCS. In 2015, some studies were conducted on the effect of IOT as an admixture on cement modified BCS. The sieve analysis showed that the percentage of fines decreased with rise in tailings content from 74.2% to 65.5% with 0% cement and 4% IOT treatment, and this trend was observed for other mixes too. BCS, after modification with cement-IOT blends, falls within the range of well-graded soil. The increased tailing content also decreased the cation exchange capacity of the mix. The modified soil showed increased maximum dry density (MDD) and decreased optimum moisture content (OMC) with higher cement-IOT content. The shear strength decreased to a minimum value till 6% tailings content before increasing. The liquid limit dropped with a rise in cement-IOT content. IOT also decreased the plastic limit the plasticity index of treated soil increased with increased cement-IOT content. The workability of the soil also improved, it became more friable with the addition of cement-IOT mix. The micro analysis of the modified BCS showed decreased cohesion as well as variation in fabric orientation with time. The addition of IOT helped to reduce amount of cement used for the treatment of BCS. It also proved to be an efficient disposal method for IOT (Osinubi et al. 2015).

Even though IOT was able to improve the geotechnical properties of cement modified BCS, it was still not able to meet the parameters of a standard subbase material in road construction. It was in this scenario, the soil improving abilities of lime-IOT blends was studied. Strength test, durability test, compaction test, etc., were carried out. UCS, CBR and resistance to strength loss gave peak values at a blend of 8% lime and 8% IOT content. This combination met the parameters of a standard subbase materials used for construction of light traffic roads (Etim et al. 2017).

IOT is also being mixed with laterite soil for the preparation of subgrade material for road construction. Lime and cement is used to stabilize this laterite-tailing mixture. A study was conducted in Nigeria using the locally available laterite soil and IOT from a mining company. The soil belonged to SC group of USCS (Unified Soil Classification System) and had a liquid limit of 53.5% and 31.4% plasticity index. The soil quality was improved and became suitable for subbase construction with the addition of tailings. This was due to the decrease in fine fraction and atterberg limits of the soil. When the laterite-mine tailings mixture was treated with lime-cement binder, the MDD of the soil increased and OMC decreased. The addition of both tailings and binder increased CBR value of the laterite soil. The addition of binder also reduced the leaching levels except in the case of barium and chromium (Ojuri et al. 2017). In fact, the laterite properties vary from place to place widely. A study carried out on the laterite waste collected from western Indian coast revealed that it can be replaced with fine aggregates in construction industry without compromising on any strength properties (Ram Chandar et al. 2016a). Recently, the copper mine tailings are being used as a replacement for road base material through geopolymerization (Manjarrez and Zhang 2018).

2.9 USE OF MINE WASTE FOR OTHER APPLICATIONS

2.9.1 Thermal Insulation

The combination of mine tailings and recycled tyre crumbs can be used as lightweight filler material with improved thermal insulation. The bulk density and thermal properties of the mixture were found to depend on the water content, mixing ratio of mine tailings and tyre crumbs, compactive effort and size of tyre crumbs. The mine tailings should be of non-acidic nature. Lee et al. (2015) studied the use of tyre crumbs of two different particle sizes. When the replacement ratio of tyre crumbs was increased from 0 to 0.4 at a water content of 15%, the thermal conductivity decreased by 66% and bulk density by 38%. These conditions were found suitable for structural fills. The thermal conductivity was found to be sensitive to water content up to the critical water content; after that it was somewhat independent. A higher compactive effort causes increased thermal conductivity. For 0.2 mixing ratio, the thermal conductivity of the mixture with small tire crumbs (D_{50} = 0.46 mm) was 3% to 12% less than that of bigger tire crumbs (D_{50} = 2.1 mm). This can be attributed to the high air entrainment in mixture containing small tyre crumbs (Lee et al. 2015).

2.9.2 Prevention of Leakage through Holes in Geomembranes

Geomembranes may leak due to various reasons such hydraulic conductivity and hydraulic gradient transmissivity of the soil membrane interface, number and size of holes, etc. The studies conducted on the effect of tailings sand samples collected from the Highland Valley Copper mine facility in Kamloops, B.C., Canada on two geomembranes (1-mm-thick linear low-density polyethylene (LLDPE) geomembrane, 2-mm-thick high-density polyethylene (HDPE) geomembrane) showed reduction in leakage through small holes in the membrane. The leakage through 10 to 20 mm diameter holes remained unchanged even after the modification. For a 1.5 mm diameter hole, flow was reduced by 3 times. Further, the study revealed that the leakage decreased with decrease in hydraulic conductivity (kT) of the tailings (Kerry Rowe et al. 2018).

2.9.3 Water Balance Covers

Water balance covers (WBC) are used to reduce the water and oxygen seeping into the underlying wastes through water retention, while providing erosion and slope failure. Gorakhki Mohammad and Bareither Christopher (2017) conducted studies to assess the favourable reuse of mine tailing-waste rock mix in WBCs. Similar simulations were carried out to understand the hydrologic behaviour of actual WBC and modified WBC. In the hydrologic modelling, three conditions were evaluated where the actual storage layer was replaced with a layer of mine waste: (1) 1.22-m-thick layers of pure mine tailings (i.e., gold, coal, copper, and oil sand tailings), (2) 1.22-m-thick layers of mine tailings-waste rock mix and (3) mine tailing-waste rock mix storage layer with thickness yielding results comparable to actual cover. For tailings having higher clay content and lower hydraulic conductivity, the percolation rates were found to be low. With an increase in hydraulic conductivity of the storage layer, evapotranspiration and percolation increased, while the runoff decreased. Application of mine tailing-waste rock mix in WBCs enhances the sustainability of mine through reduction in costs associated with waste disposal.

2.9.4 Structural Fill

Embankments, backfills, trenches and all other construction phases which require filling are potential areas where mine tailings can be used. A successful pilot project was carried out using bauxite residue for the construction of road embankment (Kehagia 2010). The embankment was designed 75 m long, 3 m high with crown width of 8 m. The 75 m long embankment consisted of three sections each with 25 m long. Section I was made of natural soil of A – 4 group, section II was a mixture of bauxite residue (40%) and soil of A – 1 group (60%) and section III a mixture of bauxite residue with fly ash (4%). These different sections were done for comparison study. Specimens of the constituent materials of the three sections were collected and tested to find mechanical strength. It was observed that excellent workability and performance of bauxite residue material was recorded throughout the construction. Soil deformation problems due to insufficient compaction did not occur. After long operational period under traffic (15–20 trucks per day) no disintegration appeared on the body of the embankment. The highest percentage of bauxite tailing used was 60%.

2.10 SOME SUGGESTIONS FOR EFFECTIVE USE OF MINE WASTE

Attempt should be made as far as possible to reduce mine waste and tailings at generation site itself, available tailings should be used for some constructive purpose and the remaining should be disposed in an environmentally and effective way. A few suggestions are:

1. In many developing countries, still there are no stringent rules to handle the mine waste and tailings. In the mine planning stage itself, it should be made mandatory to include waste disposal arrangement, a chapter to be introduced in feasibility report itself, showing the long-term and short-term solutions for waste disposal.
2. Based on the physico-mechanical properties carried out during exploration stage, the mine waste to be separated keeping in view their future use. In general, the top fertile soil will be stored separately to be used for restoration and reclamation purpose. The hard rock can be used for pavements/construction purpose and the soft to medium can be used for back filling.
   a. Generally, mines are located in remote areas, where there will be a need to develop the basic infrastructure like roads and buildings. So, it should be made mandatory to use the waste rock as well as tailings to use in road construction. This will reduce the road construction cost, problem of depletion of natural sand and aggregates also can be avoided, and the

mine management can save lot of land under waste dumps and can avoid their stabilization problems.

b. Many research studies have proved that mine tailings can be used in making bricks, which are used for construction purpose with some additives to reduce density. Small-scale brick manufacturing plants should be established at mine site under different heads like Corporate Social Responsibility (SCR) and use of such bricks should be encouraged. All the housing projects sponsored by the Government within 100 km radius from the mine site; use of brick made from mine tailings should be made compulsory, which will reduce the financial burden on the Government to a great extent.
c. Medium to soft rock may not be useful in construction of roads and buildings, such waste to be backfilled in mined out area, using different methods. As this is an additional cost for the mine management, the concerned agency should compensate the additional cost in royalty, CSR, etc., which would be a great help to not only mining industry but for the environment.
d. Back filling can be done even in surface mines, but it requires a systematic study to maintain the stability and may require some additives like cement, bentonite, etc. In some countries, many old abandoned mines were converted into picnic spots by back filling to certain extent.

3. Sometimes, the soft waste rock will have good nutrients, as the low content mineral cannot be extracted economically. In such cases, the waste can be spread over barren lands so that some special species can be planted to make it a green zone which will be very useful for environment.
4. Mine tailings can also be used effectively to back fill into underground stopes using hydraulic stowing system. In general when there is strata control problems like cut and fill method of stoping, then only such back filling is being done. But it should be made mandatory to avoid the storage problems in the tailings pond on the surface.
5. Still excessive waste/ tailings available at mine site, instead of storing at mine site permanently, those can be filled in low lying areas systematically to use the land for future use.
6. Still for additional use, latest techniques like geopolymerization, bio-geo engineering applications to be explored further.
7. There is a need of coordination among different agencies, for example in India, Indian Bureau of Mines, Director General of Mines Safety, Pollution Control Board, Ministry of Mines, National Highway Authority of India, scientific and research organizations, etc.

2.11 SUMMARY

The need for minerals is on the rise every year, so many mining projects are under expansion to increase the production, which simultaneously increases waste and tailings also. Hence, the waste management systems associated with mining should also undergo significant development to safely dispose the mine waste and tailings generated.

In civil engineering projects, recycled mine waste and tailings can be used in the manufacture of building materials, backfilling, ground improvement, soil stabilization and other miscellaneous applications. All the geotechnical applications of mine tailings mentioned above are environmental friendly methods to dispose the mine waste and tailings which reduces the environmental footprint left behind by the mining industry. Even mine waste handling is expensive in some cases; still it is essential to give a better life for the future generations on this planet.

REFERENCES

Abdulrahman, H. S. 2015. Potential use of iron ore tailings in sand crete block making. *International Journal of Research in Engineering and Technology*, 4(4), 409–414.

Amjad, Tariq and Ernest, K. Yanful. 2013. A review of binders used in cemented paste tailings for underground and surface disposal practices. *Journal of Environmental Management*, 131, https://doi.org/10.1016/j.jenvman.2013.09.039.

Barry, A. W. and James, A. F. 2016. Chapter 16—Tailings disposal. *Wills' Mineral Processing Technology* (8th Edition) https://doi.org/10.1016/B978-0-08-097053-0.00016-9

Benzaazoua, Mostafa, Bussière, Bruno, Demers, Isabelle, Aubertin, Michel, Fried, Eliane and Blier, Annie. 2008. Integrated mine tailings management by combining environmental desulphurization and cemented paste backfill: Application to mine Doyon, Quebec, Canada. *Minerals Engineering*, 21, 330–340. 10.1016/j.mineng.2007.11.012.

Carrasco, E. V. M. Mantilla, J. N. R. Esposito, T. and Moreira, L. E. 2013. Compression performance of walls of interlocking bricks made of iron ore byproducts and cement. *International Journal of Civil & Environmental Engineering*, 13(3), 56–62.

Charles, C., Jingyi, Zhua, Wayne, Jensena and Maher, Tadrosb. 2005. High-percentage replacement of cement with fly ash for reinforced. *Cement and Concrete Research*, 1088–1091.

Das, S. K. Kumar, S. and Ramachandrarao, P. 2000. Exploitation of iron ore tailing for the development of ceramic tiles. *Waste Management*, 20(8), 725–729.

Dhir, R. K. Munday, J. G. L. and Ong, L. T. 1986. Investigation on engineering properties of opc/pulverized fly ash concrete-deformation properties. *The Structural Engineer*, 64B2, 26–42.

Duxson-Loya, E. I. Allouche, E. N. and Vaidya, S. 2007. Mechanical properties of fly ash based geopolymer concrete. *ACI Material Journal*, 108, 300–306.

Erol, Yilmaz, Tikou, Belem and Mostafa, Benzaazoua. 2014. Effects of curing and stress conditions on hydromechanical, geotechnical and geochemical properties of cemented paste backfill. *Engineering Geology*, 168. https://doi.org/10.1016/j.enggeo.2013.10.024

Etim, Roland, Eberemu, Adria Oshioname and Osinubi, K. 2017. Stabilization of black cotton soil with lime and iron ore tailings admixture. *Transportation Geotechnics*, 10. 10.1016/j.trgeo.2017.01.002.

Fernandez-Jimenez, A. and Palomo, A. 2003. Characterisation of fly ash: Potential reactivity as alkaline cements. *Fuel*, 82(18), 2259–2265.

Francis, A. Kuranchie, Sanjay, K. Shukla and Daryoush, Habibi. 2014. Utilization of iron ore tailings for the production of geo polymer bricks. *International Journal of Mining, Reclamation and Environment*, 30(2), 92–114.

Giri Babu, S.V. and Krishnaiah, S. 2018. Manufacturing of eco-friendly brick: A critical review. *International Journal of Computational Engineering Research (IJCER)*, 8, 24–32.

Gorakhki Mohammad, H. and Bareither Christopher, A. 2017. Sustainable reuse of mine tailings and waste rock as water-balance covers. *Minerals* 7(7), 128.

Hammond, A.A. 1988. Mining and quarrying wastes: A critical review. *Engineering Geology*, 25 (1), 17–31.

Huang, X., Ravi, R. and Li, Victor. 2013. Feasibility study of developing green ECC using iron ore tailings powder as cement replacement. *Journal of Materials in Civil Engineering*, 25(7), 923–931.

Jemish, M. Sujeet Bharti, V. Aruna, M. and Vardhan, Harsha. 2014. Utilization of mining wastes in manufacturing of bricks. *Conference: National Seminar on Recent Trends in Mechanized Mining*, Kothagudem, Telangana, 27–28.

Joni, Safaat Adiansyah, Michele, Rosano, Sue, Vink and Greg, Keir. 2015. A framework for a sustainable approach to mine tailings management: Disposal strategies. *Journal of Cleaner Production*, 108, Part A, https://doi.org/10.1016/j.jclepro.2015.07.139

Kehagia, F. 2010. A successful pilot project demonstrating re-use potential of bauxite residue in embankment construction. *Resources, Conservation and Recycling*, 54(7), 417–421.

Kerry Rowe, R., Joshi, Prabeen, Brachman, R.W.I. and McLeod, H. 2018. Leakage through holes in geomembranes below saturated tailings. *Journal of Geotechnical and Geoenvironmental Engineering*, 144, 07018001. https://doi.org/10.1061/(ASCE)GT.1943-5606.0001814

Kesavulu, N. C. 2009. *Textbook of Engineering Geology*, MacMillan, New Delhi.

Kshitija, K. Vikrant, S. Mujahed, S. Mahesh, K. Mahendra, C. and Sagar, S. 2015. Use of iron ore tailings as a construction material. *International Journal of Conceptions on Mechanical and Civil Engineering*, 1(2), 1–21.

Kumar, S. Kumar, R. and Amitava, B. 2006. Innovative methodologies for the utilization waste from metallurgical and allied industries. *Resources Conservation and Recycling*, 48(4), 301–314.

Kumar, S. B.N. 2014. Utilzation of iron ore tailings as replacement to fine aggregates in cement concrete pavements. *International Journal of Research in Engineering and Technology*, 3(7), 369–376.

Kurama, H. Topçu, I. B. and Karakurt, C. 2009. Properties of the autoclaved aerated concrete produced from coal bottom ash. *Journal of Materials Processing Technology*, 209(2), 767–773.

Lee, J. K., Julie, Q. Shang and Sangseom, Jeong. 2015. Thermal conductivity of compacted fill with mine tailings and recycled tire particles. *Soils and Foundations*, 55(6).

Likith, N. P., Manjunatha, Siddesh, S. S., Manjunath, K. M., Shivakumar, Hadimani and Shivkumar, B. (2017). Manufacturing of building blocks by utilizing of iron ore tailings. *International Journal of Engineering Science and Computing*, 7(5), 12274–12277.

Lu, Zengxiang and Cai, Meifeng. 2012. Disposal methods on solid wastes from mines in transition from open-pit to underground mining. *Procedia Environmental Sciences*, 16. 715–721. 10.1016/j.proenv.2012.10.098

Ma, B. Guo, Cai, L. Xiong, Li, X. Guo and Jian, S. Wei. 2015. Utilization of iron tailings as substitute in autoclaved aerated concrete: Physico-mechanical and microstructure of hydration products. *Journal of Cleaner Production*, 127, 162–171.

Maharana, J. K. and Patel, A.K. 2013. Physico-chemical characterization and mine soil genesis in age series coal mine overburden spoil in chromosequence in a dry tropical environment. *Phylogenetics & Evolutionary Biology*, 1(1), 1–7.

Mangalpady, Aruna. 2012. Utilization of iron ore tailings in manufacturing of paving blocks for eco-friendly mining, IC-GWBT2012, Ahmad Dahlan University.

Manjarrez, Lino and Zhang, Lianyang. 2018. Utilization of copper mine tailings as road base construction material through geopolymerization. *Journal of Materials in Civil Engineering*. DOI: 10.1061/(ASCE)MT.1943-5533.0002397.

Mendes, Beatryz C., Pedroto, Leonardo G., Fontes, Mauricio P.F., Ribeiro, Jose Carlos L., Vieira, Carlos M.F., Pacheco, Anderson A. and de Azevedo, Afonso R.G. 2019. Technical and environmental assessment of the incorporation of iron ore tailings in construction clay bricks. *Construction and Building Materials*, 227.

Mindess, S. Young, J. F. and Darwin, D. 2003. *Concrete* (2nd Edition), Prentice-Hall Inc., Englewood Cliffs, NJ.

Mirza, W. H. and Al-Noury, S. I. 1986. Utilisation of Saudi sands for aerated concrete production. *International Journal of Cement Composites and Lightweight*, 8(2), 81–85.

Mostafa, N. Y. 2005. Influence of air-cooled slag on physicochemical properties of autoclaved aerated concrete. *Cement and Concrete Research*, 35(7), 1349–1357.

Nagaraj, H.B. Rajesh, A. and Sravan, M.V. 2016. Influence of soil gradation proportion and combination of admixtures on the properties and durability of CSEBS. *Construction and Building Materials*, 110, 135–144.

Osinubi, K., Yohanna, Paul and Eberemu, Adrian Oshioname. 2015. Cement modification of tropical black clay using iron ore tailings as admixture. *Transportation Geotechnics*, 5. 10.1016/j.trgeo.2015.10.001.

Palomo, A. Grutzeck, M. W. and Blanco, M. T. 1999. Alkali-activated fly ashes: A cement for the future. *Cement and Concrete Research*, 29(8), 1323–1329.

Prahallada, M. C. and Shanmuka, K. N. 2014. Stabilized iron ore tailings block an – environmental friendly construction material. *International Journal of IT, Engineering and Applied Sciences Research*, 3(4), 2319–4413.

Prashant, S., Ghosh, C. N. and Mandal, P. K. 2010. Use of crushed and washed overburden for stowing in underground mines: A case study. *Journal of Mines Metals and Fuels*, 7–12.

Prem Kumar, W. P. Ananthya, M. B. and Vijay, K. 2014. Effect of replacing sand by iron ore tailings on the compressive strength of concrete and flexural strength of reinforced concrete blocks. *International Journal on Research in Engineering and Technology*, 3(7), 1374–1376.

Punmia, B. C. and Jain, A. K. 2005. *Soil Mechanics and Foundations*. Laxmi Publications, New Delhi, 98–101.

Rai, A. K., Paul, B. and Singh, G. 2010. A study on the bulk density and its effect on the growth of selected grasses. *International Journal of Environmental Sciences*, 1(4), 677–684.

Ram Chandar, K., Raghunandan, M. E. and Manjunath, B. 2016a. Partial replacement of fine aggregates with laterite in GGBS-blended-concrete *Journal of Advances in Concrete Construction*, 4, (3), 221–230, http://dx.doi.org/10.12989/acc.2016.4.3.221

Ravikumar, C. M., Kumar, A., Prashanth, M. H. and Reddy, D. V. 2012. Experimental studies on iron ore tailings based interlocking paver blocks, *International Journal of Earth Sciences*, 5(3), 501–504.

Shetty. 2005. *Concrete Techonology*. S. Chand & Company Ltd., New Delhi.

Shreekant, R. L., Aruna, M. and Harsha, Vardhan. 2016. Utilization of iron ore waste in brick making for the construction industry. *International Journal of Earth Sciences and Engineering*, 9(2), 182–195.

Ugama, T. I. and Ejeh, S. P. 2014. Iron ore tailings as fine aggregate in mortar used for Masonry. *International Journal of Advances in Engineering and Technology*, 7(4), 1170–1178.

Ullas, S.N., Venkatarama Reddy, B.V. and Nanjunda Rao, K.S. 2010. Characteristics of masonry units from iron ore tailings. *International Conference on Sustainable Built Environment, Kandy*, 108–114. https://minerals.usgs.gov/minerals/pubs/commodity/iron_ore/mcs-2017-feore.pdf. Accessed on 6th September 2018.

Wang, C. L. Ni, W. Zhang, S. Q. Wang, S. Gai, G. S. and Wang, W. K. 2016. Preparation and properties of autoclaved aerated concrete using coal gangue and iron ore tailings. *Construction and Building Materials*, 104, 109–115.

Wei, Sun, Hongjiang, Wang and Kepeng, Hou. 2018. Control of waste rock-tailings paste backfill for active mining subsidence areas. *Journal of Cleaner Production*, 171. https://doi.org/10.1016/j.jclepro.2017.09.253

Yaseen, Sayar, Amit, Pal, Siddharth, Singh and Idrees, Yousuf Dar. 2012. A study of physico-chemical characteristics of overburden. *Global Journal of Science Frontier Research*, 7–13.

Yin, Shenghua, Shao, Yajian, Wu, Aixiang, Wang, Yiming and Chen, Xun. 2018. Expansion and strength properties of cemented backfill using sulphidic mill tailings. *Construction and Building Materials*, 165, 450–455. https://doi.org/10.1016/j.conbuildmat.2018.01.005

Yisa, G. L. Akanbi, D. O. Agbonkhese, O. Ahmed, M. K. and Sani, J. E. 2016. Effect of iron ore tailing on compressive strength of manufactured laterite bricks and its reliability estimate. *Civil and Environmental Research*, 8(8), 49–58.

Yongliang, Chen, Yimin, Zhang, Tiejun, Chen, Yunliang, Zhao and Shenxu, Bao. 2010. Preparation of eco-friendly construction bricks from hematite tailings. *Construction and Building Materials*, 25, 2107–2111.

Zhao, S. Fan, J. and Sun, W. 2014. Utilization of iron ore tailings as fine aggregate in ultra-high performance concrete. *Construction and Building Materials*, 50, 540–548.

Zheng, Juanrong, Zhu, Yalan and Zhao, Zhenbo. 2016. Utilization of limestone powder and water-reducing admixture in cemented paste backfill of coarse copper mine tailings. *Construction and Building Materials*, 124. 31–36. 10.1016/j.conbuildmat.2016.07.055. https://shodhganga.inflibnet.ac.in/bitstream/10603/152177/16/9.%20chapter-1.pdf. Accessed on 20th September 2017

Zhengfu, Bian, Inyang, Hilary I., Daniels, John L., Otto, Frank and Struthers, Sue. 2010. Environmental issues from coal mining and their solutions. *Mining Science and Technology* (China), 20(2), 215–223.

IS CODES

IS 3812-1 (2003): "Specification for pulverized fuel ash, Part 1: For use as pozzolana in cement, cement mortar and concrete." *Bureau of Indian Standards*, New Delhi.

IS: 10262-2009. "Concrete mix proportioning – guidelines." *Bureau of Indian Standards*, New Delhi.

IS: 2386-1963. "Methods of tests for aggregates for concrete." *Bureau of Indian Standards*, New Delhi.

IS: 383-1970. "Specifications for coarse and fine aggregates from natural sources of concrete." *Bureau of Indian Standards*, New Delhi.

IS: 456-2000. "Code of practice for plain and reinforced concrete." *Bureau of Indian Standards*, New Delhi.

IS: 516-1959. "Methods of test for strength of concrete." *Bureau of Indian Standards*, New Delhi.

IS: 5816-1999. "Splitting tensile strength of concrete – method of test." *Bureau of Indian Standards*, New Delhi.

BIBLIOGRAPHY

Chen, Q. S., Zhang, Q. L., Fourie, A., Chen, L. X. and Qi, C. C. 2017a. Experimental investigation on the strength characteristics of cement paste backfill in a similar stope model and its mechanism. *Construction and Building Materials*, 154, 34–43. DOI: 10.1016/j.conbuildmat.2017.07.142.

Chen, Qiusong, Zhang, Qinli, Fourie, Andy and Xin, Chen. 2017b. Utilization of phosphogypsum and phosphate tailings for cemented paste backfill. *Journal of Environmental Management*, 201, https://doi.org/10.1016/j.jenvman.2017.06.027.

Gayana, B. C. and Ram Chandar, K. 2018. Sustainable use of mine waste and tailings with suitable admixture as aggregates in concrete pavements-A review. *Journal of Advanced Concrete Technology*, 6(3), 221–243.

Hu, Li-Ming, Wu, Hui, Zhang, Lin, Zhang, Pengwei and Wen, Qingbo. 2016. Study on static and dynamic strength characteristics of tailings silty sand and its engineering application. *Journal of Materials in Civil Engineering*, 29. 04016220. 10.1061/(ASCE)MT.1943-5533.0001736.

Ojuri, O. O., Adavi, A. A. and Oluwatuyi, O. E.. 2017. Geotechnical and environmental evaluation of lime–cement stabilized soil–mine tailing mixtures for highway construction. *Transportation Geotechnics* 10, 1–12.

Panchal, Sandeep, Deb, Debasis and Sreenivas, T. 2018. Mill tailings based composites as paste backfill in mines of U-bearing dolomitic limestone ore. *Journal of Rock Mechanics and Geotechnical Engineering*, 10. 10.1016/j.jrmge.2017.08.004.

Ram Chandar, K. Deo, S. N. and Baliga, A. J. 2016. Prediction of Bond's Work Index from field measurable rock properties. *International Journal of Mineral Processing*, 157(2016),134–144.

Ramanaidou, E. R. and Wells, M. A. 2014. Sedimentary hosted iron ores. In *Treatise on Geochemistry* (2nd Edition), Edited by: H.D. Holland and K.K. Turekian, Oxford: Elsevier, 313–355.

Sun, J. S., Dou, Y. M., Chen, Z. X. and Yang, C. F. (2011), "Experimental study on the performances of cement stabilized iron ore tailing gravel in highway application. *Journal of Applied Mechanics*,97–98, 425–428.

3 Utilization of Coal Mine Waste in Concrete

Sainath Vanakuri and Ram Chandar Karra

CONTENTS

3.1 INTRODUCTION

In general, coal mine overburden comprises run-of-mine waste, coarse grained washery discard, fine grained discard and slurry, rejects and tailings. Petro graphically coal mine overburden consists of argillaceous and arenaceous rocks represents mainly by mudstone, siltstone and sandstone with admixture of coal and coal shale (Twardowska et al. 2004). Mineralogically, shale is mainly made of montmorillonite (Kesavulu 2009). Montmorillonite is actually highly active clay (Punmia and Jain 2005). As using of clay in concrete is not possible, using of shale is also not possible.

On the other hand, sandstone contains more than 90% of the particles as sand and possibilities are there to use sandstone in concrete (Kesavulu 2009).

DOI: 10.1201/9781003268499-3

3.2 USES OF COAL MINE WASTE

Many studies were carried out in India about coal mine waste. A study on the coal mine overburden, where crushed over burden is used for stowing purpose stated that there is 93% of the sand content present in the over burden and specific gravity of the over burden was found to be on an average 2.53. Compressibility characteristics were also found and it was concluded that it is useful for stowing in underground mines, if the material washes through 0.15 mm sieve (Prashant et al. 2010). A study on Jharia coal mining fields revealed that 98% of the particles are retained on 75 μm and higher sieve sizes (Rai et al. 2010). A study on Raniganj coal field concluded that there was 96% of sand content in the over burden and the presence of clay and silt particles was very low. In this material the presence of nitrogen, potassium and phosphorus contents are very less and the soil is not useful for plantation (Sayar et al. 2012). A study carried out on the coal mine overburden of the Basundhara (west) open cast colliery reveals the sand content in the samples are more than 80%. The study suggested that the fresh mine spoil to attain the soil features of native forest soil through the process of reclamation shall take 28 years, provided the spoil habitat is not subjected to any other interferences like erosion, vegetational degradation, etc. (Maharana and Patel 2013).

3.2.1 Sandstone as Replacement for Fine Aggregate

A study on replacement of fine aggregate with crushed sandstone as 0%, 20%, 40% and 60 % compressive strength on HPC using crushed sandstone sand has been observed under water curing. It was concluded that there is high potential for utilization of crushed sandstone sand in HPC production (Kumar et al. 2005). Crushed sandstone was used as fine aggregate in the concrete and the results showed that it is a good replacement for fine aggregate (Fontanini et al. 2013). The replacement of fine aggregate with coal mine overburden in concrete paving block showed the maximum strength at 50% and beyond that it reduces the strength of the paving block (Santos et al. 2013).

3.2.2 Sandstone as Replacement for Coarse Aggregate

Beams made with coarse aggregate as sandstone, due to lower stiffness of sandstone aggregates, resulted in excessive deflection under service loads but it was within acceptable limit (Kumar et al. 2007). Five different aggregate types (gabbro, basalt, quartzite, limestone and sandstone) were used to produce high strength concrete containing silica fume and among them the lowest strength was obtained by sandstone (Kiliç et al. 2007). When coarse aggregate was replaced by sandstone aggregate, the results obtained showed that subarkose–arkose, sublitharenite–litharenite and arkose aggregates which have clay cement caused approximately a 40–50% reduction in concrete strength when compared to subarkose, quartz sandstone and arkose aggregates which have carbonate cement because these aggregates result in weaker bonding between aggregate and cement than others. Five different aggregate types such as, gabbro, basalt, quartzite, limestone and sandstone were to produce high strength concrete. It was concluded that gabbro concrete showed the highest compressive and, while sandstone showed the lowest compressive strength (Mucteba 2012).

3.2.3 Fly Ash as a Replacement of Cement

Fly ash is finely divided residue resulting from the combustion of powdered coal and transported by the flue gases and collected by electro static precipitator. Fly ash is mostly used Pozzolanic material all over the world (Shetty 2005).Manufacturers of reinforced concrete products commonly limit the quantity of fly ash to 25% or less by weight (Charles et al. 2005).

Unused fly ash is usually disposed into landfills contributing to soil, water and air pollution (Palomo et al. 1999; Duxson-Loya et al. 2007). As a product of the burning of coal, its type is determined by the type of coal. The anthracite and bituminous coals produce low calcium fly ash which possesses truly pozzolanic properties due to the high content of silica, while the lignite or sub bituminous coals produce high calcium fly ash which is both a cementitious and pozzolanic material it has lower silica and alumina but higher CaO content (Dhir et al. 1986).

There are two ways that fly ash can be used. One way is to mix certain amount of fly ash with cement clinker to produce Portland pozzolana cement and the second way is to use the fly ash as admixture at the time of making concrete at the site of work (Shetty 2005). ASTM broadly classifies fly ash into two types. They are class F and class C. Class F is the fly ash normally produced by burning anthracite or bituminous coal. Class C is the fly ash produced by burning lignite or sub-bituminous coal (ASTM C618).

A high level of SO_3 in concrete can lead to volume instability and thus loss of durability, due to the formation of ettringite. Therefore IS 3812- part- 1(2003) allows for a maximum of 3% of total sulphur as SO_3 for fly ash to be used for concrete as binder. Finesses of the fly ash play an important role in the development of the mechanical strength. The low-calcium fly ash should have the percentage of unburnt carbon (LOI) more than Fe_2O_3 content should be less than 10%, should have low CaO content and the content of reactive silica should be between 40%–50% (Fernandez-Jimenez and Palomo 2003).

In order to assess the suitability of coal mine over burden, systematic investigations are carried out.

3.3 EXPERIMENTAL INVESTIGATIONS

3.3.1 Sample Collection

Sandstone samples were collected from five different locations of overburden dumps of a coal mine in South India (Figure 3.1). The primary objective while collecting samples for laboratory analysis is that its composition should be representative of the conditions that exist in the field. The procedure involves the random collection of material in the site over the designated area and combining them to form a composite sample for analysis.

FIGURE 3.1 View of sandstone over burden dump.

3.3.2 Properties of Sandstone

3.3.2.1 Grain Size Distribution

Grain size distribution was done by sieve analysis, with sieves ranging from 75 μm to 4,750 μm. For wet sieve analysis, the sample was washed and passed through the sieves and dried for 24 hours. Then the weights were taken for materials retained on each sieve. Based on the values obtained, cumulative passed and cumulative retained were calculated for each sieve.

Sieve analysis was done for 5 different samples and cumulative percentage of passing through different sieves is given in Table 3.1. From Table 3.1 it can be observed that, in all the 5 samples 92% of the particles are greater than 75 μm size,

TABLE 3.1
Cumulative Percentage Passed for all Samples

	Cumulative Percentage Retained (%)				
IS Sieve Size (micron)	Sample 1	Sample 2	Sample 3	Sample 4	Sample 5
4750	1.57	1.25	3.06	2.81	1.91
2360	3.74	3.89	5.67	5.92	3.21
1180	15.58	14.25	16.39	18.53	15.33
600	41.42	42.96	41.43	38.94	39.92
300	71.80	72.53	75.76	71.92	70.97
150	89.49	90.55	87.73	88.79	91.91
75	93.82	93.32	94.62	92.98	93.15

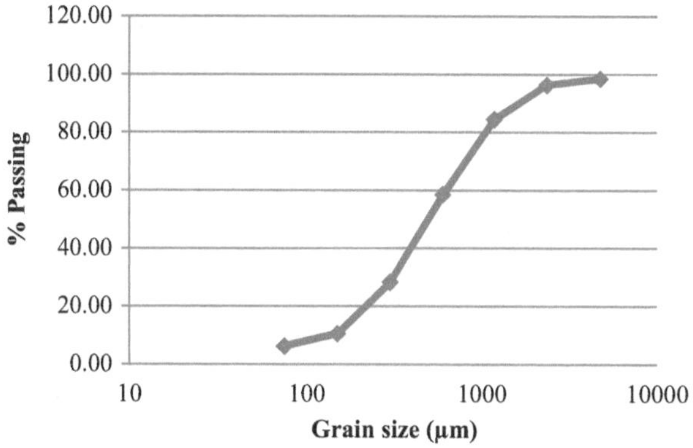

FIGURE 3.2 Grain size distribution of sandstone sample.

which is the most important property to use it as a replacement for fine aggregate. As the code suggests for crushed or artificial sand, the percentage of particles >75 µm should be more than 85%. From the table, it can be observed that more than 50% of the particles accumulated in the range of 300–1180 µm. The particles which pass through 75 µm sieve are 6.17, 6.67, 5.39, 7.01 and 6.86% respectively for each sample. The particles which pass through 75 µm can be clay, silt particles or cementitious content.

Based on the results obtained, a plot was drawn for a typical sample as shown in Figure 3.2. On X-axis log scale was taken and grain size was considered. On the Y-axis the cumulative percentage of particles passed was given. Based on this plot, the zones of sand were decided. The fineness modulus was found for the sum of cumulative percentage of particles that are retained on 150 µm and above.

Fineness modulus of samples is varying from 2.23 to 2.30, with an average value of 2.25. It indicates that the material will come under fine sand which is perfectly suitable as replacement for fine aggregates in concrete. Figure 3.3 represents the grain size distribution and from that it could be stated that all the samples have similar particle size range.

3.3.2.2 Specific Gravity and Water Absorption

Specific gravity and water absorption tests were conducted for five samples using pycnometer. Table 3.2 gives the specific gravity and water absorption for each sample. This table shows that all the samples have almost similar results, indicating that the sandstone is almost uniform and possess homogeneous properties. The specific gravity of five samples is 2.56, 2.57, 2.55, 2.57 and 2.55. There is no deviation for any samples more than 0.02 with average specific gravity value considered as 2.56 for the mix design purposes. This value is closer to conventional fine aggregate. So, the weight of the cubes won't decrease much when the fine aggregate is replaced with sandstone.

Water absorption test is continuation of specific gravity test and done with pycnometer for five samples. The water absorptions for five samples was found to be 2.21%, 2.33%, 2.24%, 2.19% and 2.28% with an average of 2.25%. The water

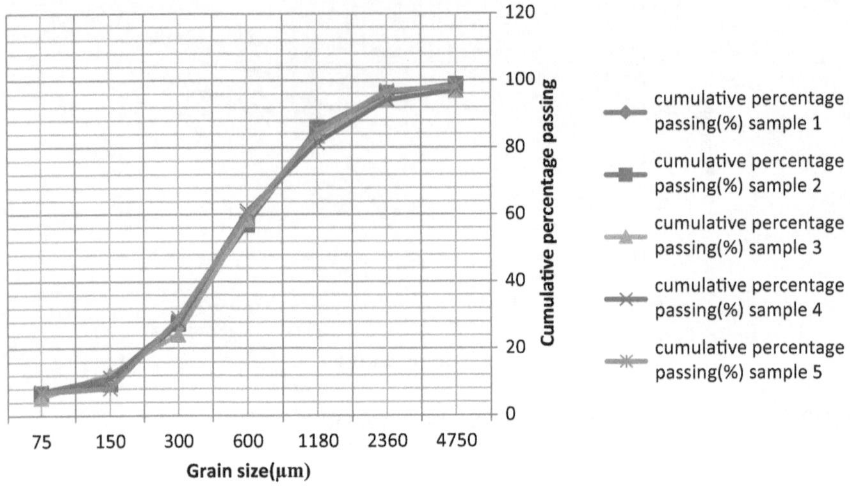

FIGURE 3.3 Particle size distributions for all samples.

TABLE 3.2
Specific Gravity and Water Absorption

Parameter	Sample 1	Sample 3	Sample 2	Sample 4	Sample 5
Weight of saturated surface-dry sample (gm)	510.23	511.16	512.49	511.2	515.17
Weight of pycnometer or gas jar containing sample and filled with distilled water (gm)	1860.38	1862.85	1860.37	1860.58	1865.85
Weight of pycnometer or gas jar filled with distilled water only (gm)	1545.25	1546.15	1544.28	1544.21	1546.87
Weight of oven-dried sample (gm)	499.21	499.53	501.26	500.26	503.71
Apparent Specific Gravity	2.71	2.73	2.71	2.72	2.73
Specific Gravity, (G)	2.56	2.57	2.55	2.57	2.57
Water Absorption (%)	2.21	2.33	2.24	2.19	2.28

absorption for Conventional fine aggregate was found to be 0.5%. So when the concrete is prepared with replacing sandstone as a replacement, the extra amount of water has to be added to prevent the concrete from the workability point of view.

3.3.2.3 Moisture Content

Moisture content was determined for all samples of sandstone. Table 3.3 specifies the moisture content calculations from the observations made.

The moisture content of samples is varying between 2.3% to 2.5% of weight of the material. So the material is easy for transportation. For the design considerations, an average value of 2.4% is considered.

TABLE 3.3
Moisture Content of Sandstone

Trial No	Cup No	Weight of Cup (gm)	Weight of Cup with Sample (gm)	Weight of Cup with Sample after Oven Drying (gm)	Weight of Water (gm)	Moisture Content $W = \frac{Ww}{Ws} \times 100$
1	52	33.35	79.8	78.67	1.13	2.49
2	56	32.65	85.98	84.67	1.31	2.52
3	62	34.12	82.19	81.08	1.11	2.36
4	57	33.96	81.76	80.64	1.12	2.40
5	66	32.87	80.23	79.19	1.04	2.25

It is to be remembered that, while designing of concrete mix this 2.4% of water content to be reduced by weight from the mix, so that the workability and strength of concrete can't be disturbed.

3.3.2.4 Scanning Electron Microscope Analysis

Scanning Electron Microscope (SEM) analysis was conducted on sandstone samples. The SEM gives two views for each sample, which means the scanning was done on two faces. Based on the plots obtained, the calculations were auto generated for elemental composition. Figure 3.4 shows the microscopic views of sandstone.

Figure 3.5 represents the auto generated plot of a typical sample of first face.

Based on the plots obtained, EDAX was calculated for the elemental composition. The elemental composition of samples is given in Table 3.4.

Elemental composition of sandstone is tested for two samples using SEM analysis added with EDAX. Along with the elements mentioned in table there is presence of materials such as magnesium in small dosage. It can be concluded that the sandstone primarily consists of oxygen, aluminium, iron, silicon and traces of titanium and

FIGURE 3.4 Microscopic views of sample 1.

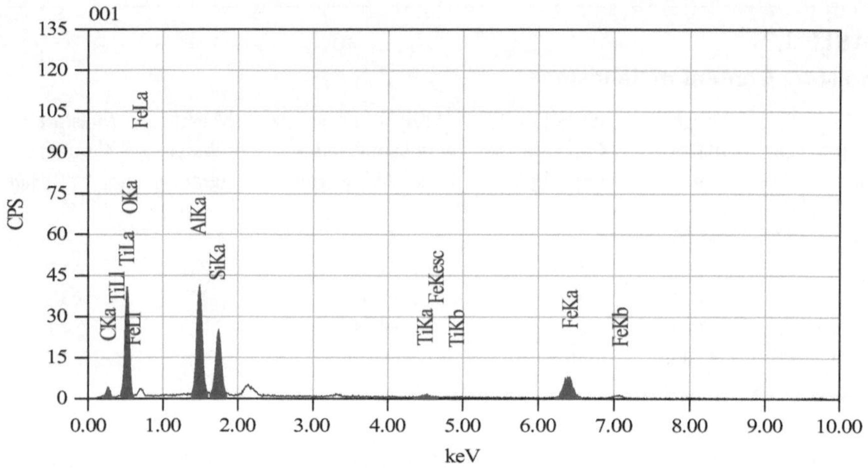

FIGURE 3.5 Plot showing elements present in sample 1.

TABLE 3.4
Elemental Composition of Sample 1 (View 2)

Element	(keV)	Mass	Error	At%	k
C	0.277	7.79	0.12	12.21	1.2446
O	0.525	52.24	0.08	61.46	62.6070
Al	1.486	23.20	0.04	16.19	21.4069
Si	1.739	13.38	0.06	8.97	10.8719
Ti	4.508	0.58	0.08	0.23	0.6602
Fe	6.398	2.81	0.95	0.95	3.2094

potassium in the form of oxides. It can be concluded that there are no traces of sulphur content in the sample. Hence, there is no threat from internal sulphate attack to concrete internally. There is no calcium content in the samples, and therefore it can resist the acid attack.

3.3.2.5 Workability of Fresh Concrete

Workability of the concrete was tested by slump cone test. Figure 3.6 shows the mould that is kept for test process and Figure 3.7 shows the flow of slump after releasing the mould. After lifting the mould, the flow was occurred. Then the height of the slump collapsed was measured and results were tabulated in Tables 3.5 and 3.6; from the Figure 3.7 it can be observed that slump is purely true.

Table 3.5 shows the slump values with varying sandstone and Table 3.6 shows the slump values with varying fly ash content keeping sandstone constant.

Workability of concrete is checked for each mix when it was casted. The workability was getting increased with increase in replacement of fine aggregate with sandstone as it is fine compared with conventional fine aggregate. In all the cases, the

FIGURE 3.6 Slump cone test.

FIGURE 3.7 Flow of slump after release.

TABLE 3.5
Slump Test Results (Sandstone)

Replacement of Sandstone (%)	Slump (mm)
0	78
25	82
50	84
75	85
100	85

TABLE 3.6
Slump Test Results (FA)

Replacement of Fly Ash (%)	Slump (mm)
0	86
10	87
20	89
30	94

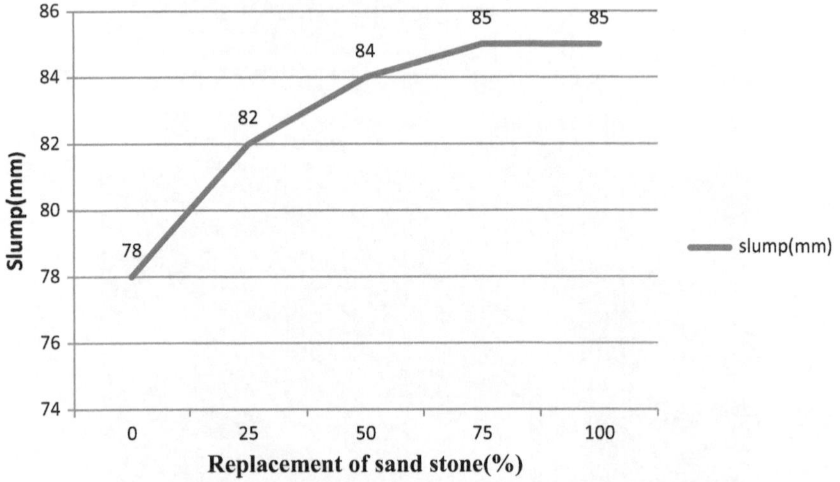

FIGURE 3.8 Variation of slump with sandstone.

type of slump was true. The variation of workability with sandstone is shown in Figure 3.8.

When fly ash was added as an admixture, the workability gets increased. The reason behind this is the fineness of fly ash. When the fineness of cementitious material increases, both consistency of cement and workability of concrete increase.

So, it is suggested that when fly ash is replaced at 30% for cement, it is better to reduce the water cement ratio by 0.05.

3.3.3 Mechanical Properties

3.3.3.1 Compressive Strength

Compressive strength of concrete was tested in laboratory for each replacement at 3, 7 and 28 days. The arrangement of compression test is shown in Figure 3.9. Figure 3.10 shows the sample subjected to compression testing along with failure mechanism.

In order to assess the effectiveness of sandstone replacement with fine aggregates, first a base mix was prepared without any sandstone. The samples were tested for 3, 7 and 28 days duration. Then the natural fine aggregate was replaced with sandstone at 25, 50, 75 and 100 percentage and the number of samples were moulded and tested for compressive strength for different curing durations.

Further, to utilize fly ash which is a waste product of coal mining, cement was partially replaced with fly ash at 10, 20 and 30 percentage keeping the sandstone percentage fixed.

3-days compressive strength: The 3-days compressive strength at 0% replacement was found to be 13.78 N/mm^2 which is quite good for M20 grade concrete.

FIGURE 3.9 Testing of cubes for compression strength.

FIGURE 3.10 Cube subjected to compression test.

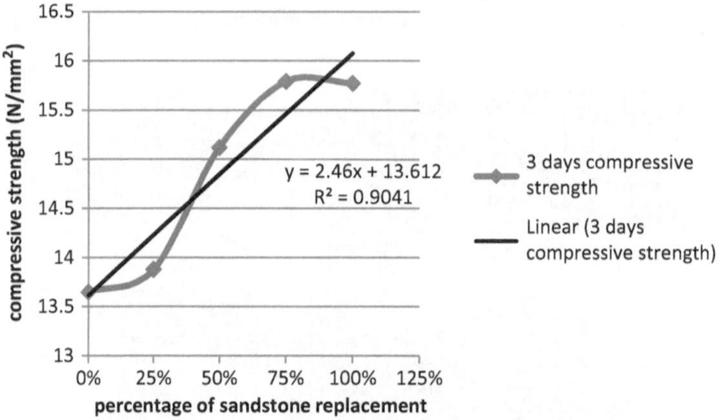

FIGURE 3.11 Variation of 3-days compressive strength with sandstone replacement.

The variation of strength with replacement of sandstone is shown in Figure 3.11. With the 25% replacement, compressive strength increased just by 0.8% with an average of 13.89 N/mm^2. From 25% to 50% replacement, strength increased by 8.07%, which is a very good variation. From 50% to 75% replacement, strength got increased by 4.44% and for 100% replacement it showed the same strength as 75% replacement. Overall, the 3-days compressive strength was increased by 14.51% for 100% replacement with sandstone.

Keeping the fine aggregate replacement as 100% with sandstone and fly ash, the replacement for cement is varied at 10%, 20% and 30%. The variation of 3-days strength with respect to variation in fly ash is shown in Figure 3.12. For 0% to 10% the strength was decreased by 4.94%. From 10% to 20%, compressive strength

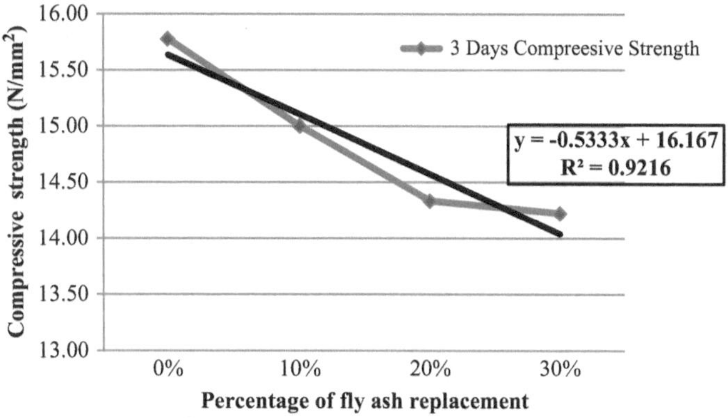

FIGURE 3.12 Variation of 3-days compressive strength with fly ash replacement.

decreased by 4.44%. From 20% to 30%, strength decreased by 0.7%. Overall, the 3-days compressive strength of concrete was decreased by 9.88% for fly ash replacement of 30% comparing with 0% fly ash.

7-days compressive strength: 7-days compressive strength of concrete with 0% replacement is found to be 17.22 N/mm^2. The variation of strength with different percentage of replacement is shown in Figure 3.13. For the replacement of 25% with sandstone, there is no increment in strength. From 25% to 50%, strength was increased by 5.16%. From 50% to 75% replacement, strength was increased by 1.21%. From 75% to 100% strength was increased by 1.25%. Overall, for 100% replacement the strength was increased by 13.4% compared with control mix, which is a very good aspect.

The variation of 7-days strength with respect to fly ash is shown in Figure 3.14. For the variations of fly ash percentages from 0% to 10% replacement, the strength was decreased by 1.83%. From 10% to 20%, compressive strength was decreased by 1.81%. From 20% to 30%, strength was decreased by 7.43%. Overall, the 7-days

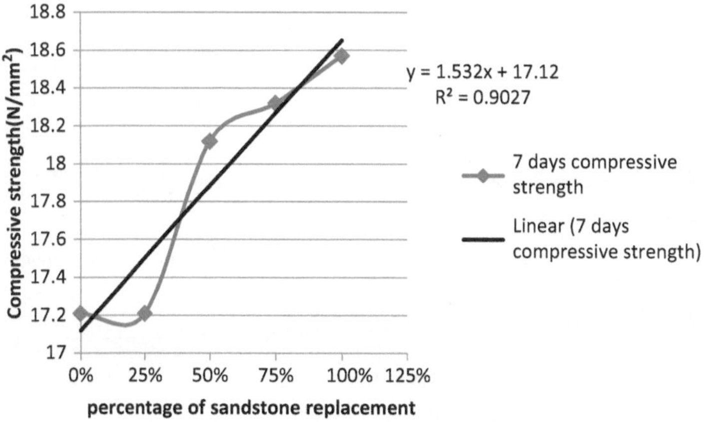

FIGURE 3.13 Variation of 7-days compressive strength with sandstone replacement.

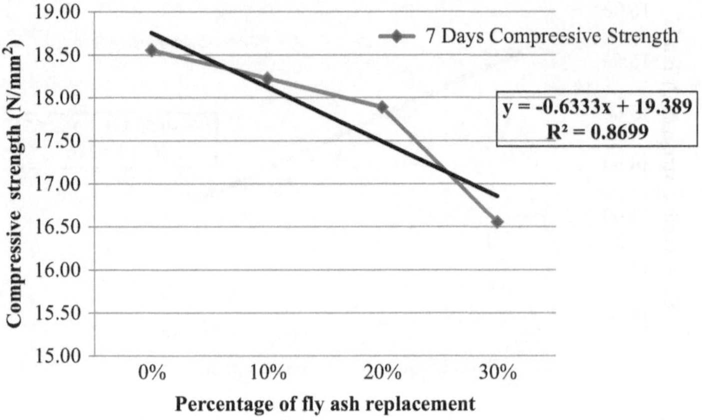

FIGURE 3.14 Variation of 7-days compressive strength with fly ash replacement.

compressive strength of concrete was decreased by 10.77% for fly ash replacement of 30% (Figure 3.14).

28-days compressive strength: Even though the 3- and 7-days compressive strength was determined, the most required strength for the concrete is after 28-days curing in general. The strength of concrete is usually talked in terms of 28-days compressive strength. 28-days strength of concrete with 0% replacement is found to be 22.56 N/mm². With the replacement of 25% with sandstone, there is 3.33% increase in strength. From 25% to 50% replacement, strength was increased by 4.28%. From 50% to 75% replacement, strength was increased by 0.36%. From 75% to 100% replacement, strength was increased by 1.6%. Overall, for 100% replacement, the strength was increased by 9.84% when compared with control mix. The strength noted at 100% replacement was 24.78 N/mm². Figure 3.15 shows the variation of compressive strength with percentage of replacement.

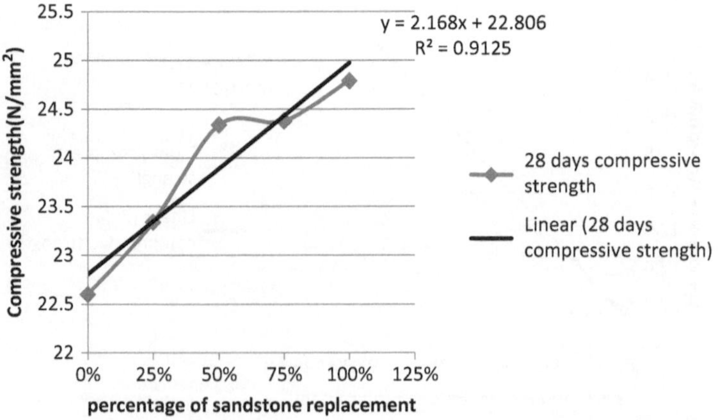

FIGURE 3.15 Variation of 28-days compressive strength with sandstone replacement.

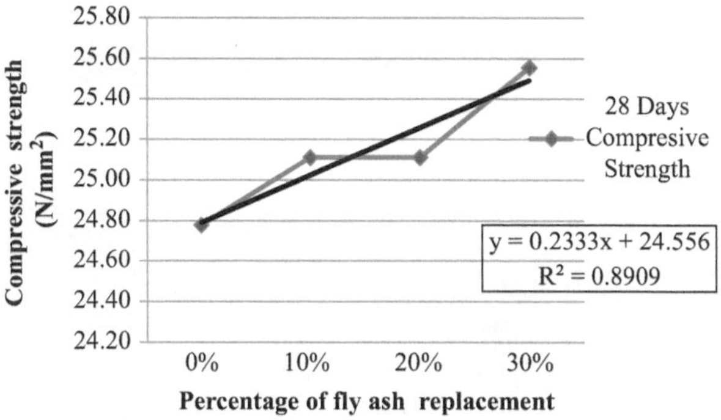

FIGURE 3.16 Variation of 28-days compressive strength with fly ash replacement.

The variation of 28-days strength with respect to fly ash is shown in Figure 3.16. For the variations of fly ash percentages from 0% to 10% replacement, the strength was increased by 1.33%. From 10% to 20% replacement, compressive strength was neither decreased nor increased. From 20% to 30%, strength was increased by 1.79%. Overall the 28-days compressive strength of concrete was increased by 3.05% for fly ash replacement of 30%.

Among the 3, 7 and 28 days compressive strength, it can be seen that in all the cases, the strength increased with the increases in replacement of fine aggregates with sandstone. The main reason could be the presence of small percentage of cementitious content that was present in coal mine overburden. It is known that small percentage of cement can cause good variation in strength.

Coming to the variation of strengths with fly ash, it was observed that the 3- and 7-days strengths were showing decreasing trend with an increase in fly ash content and increase in 28-days strength. The reason behind this is the pozzolanic action of fly ash. The addition of pozzolanic materials causes decrease in early strength and will show little higher strengths at a later stage. Because of that for fly ash replaced at 30% with cement, concrete compressive strength got increased for 28 days.

3.3.3.2 Split Tensile Strength

Split tensile strength test was conducted on moulded cylinders. For each mix, four cylinders were casted and cured for 28 days. Figure 3.17 shows split tensile strength setup and Figure 3.18 shows the failure modes of few samples.

The concrete is not usually expected to resist the direct tension because of its low tensile strength and brittle nature. However, the determination of tensile strength of concrete is necessary to determine the load at which the concrete members may crack. The cracking is a form of tension failure.

The split tensile strength at 0% replacement with sandstone (base mix) was found to be 3.18 N/mm^2 which is quite good for M20 grade concrete. The variation of strength with replacement of coal mine overburden-sandstone is shown in Figure 3.19. With the 25% replacement, strength was increased by 2.20%. From 25% to 50% replacement, strength was increased by 4.25%. From 50% to 75% replacement, strength got increased by 2.18% and for 100% replacement it showed the same

FIGURE 3.17 Arrangement for split tensile strength setup.

FIGURE 3.18 A few failed samples.

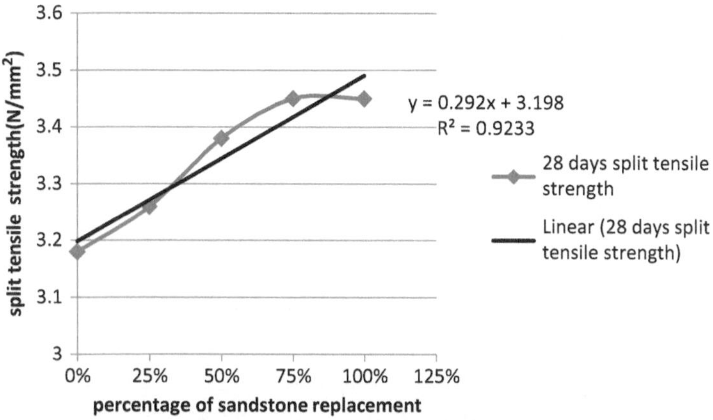

FIGURE 3.19 Variation of split tensile strength with sandstone replacement.

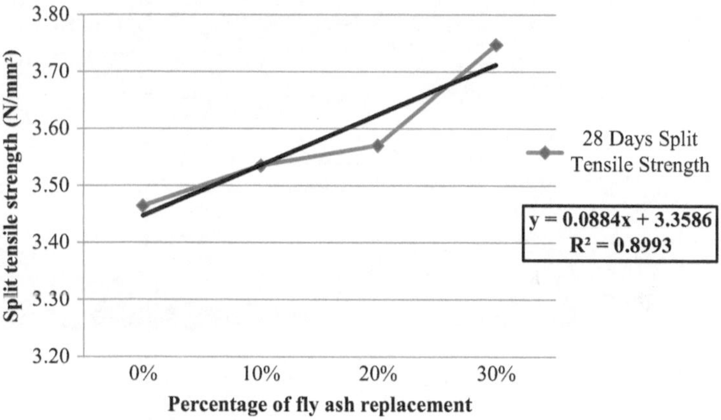

FIGURE 3.20 Variation of split tensile strength with fly ash replacement.

strength as 75% replacement. Overall the split tensile strength was increased by 8.46% for 100% replacement with sandstone.

The variation of split tensile strength with respect to fly ash replacement is shown in Figure 3.20. For the variations of fly ash percentages, from 0% to 10% replacement, the strength was increased by 1.33%. From 10% to 20% replacement, the strength was neither decreased nor increased. From 20% to 30% replacement, strength was increased by 1.79%. Overall, the split tensile strength of concrete was increased by 3.05% for fly ash replacement of 30% with a value of 3.75 N/mm².

3.3.3.3 Flexural Strength

Flexural strength test was conducted for prisms of 500×150×150 mm of concrete. For each mix, four cylinders were casted and cured for 28 days. Figure 3.21 shows the setup to conduct the test and Figure 3.22 shows the failed samples.

FIGURE 3.21 Flexural test arrangement.

The flexural strength at 0% replacement was found to be 3.00 N/mm^2 which is quite good for M20 grade concrete. The variation of strength with replacement of sandstone is shown in Figure 3.23. With the 25% replacement, strength was increased by 10%. From 25% to 50% replacement, strength was increased by 3.03%. From 50% to 75% replacement, strength got increased by 2.94% and for 100% replacement it got increased by 2.85%. Overall, the flexural strength was increased by 20% for 100% replacement with sandstone.

The variation of flexural strength with respect to fly ash replacement is shown in Figure 3.24. For the variations of fly ash percentages, from 0% to 10% replacement, the strength was marginally increased. From 10% to 20% replacement, strength was increased by 2.77%. From 20% to 30% replacement, strength was increased by 2.70%. Overall, the split tensile strength of concrete was increased by 5.55% for fly ash replacement of 30%. Overall, the flexural performance is very good at 100% replacement of fine aggregates and with admixture as fly ash.

FIGURE 3.22 failed samples in flexural text.

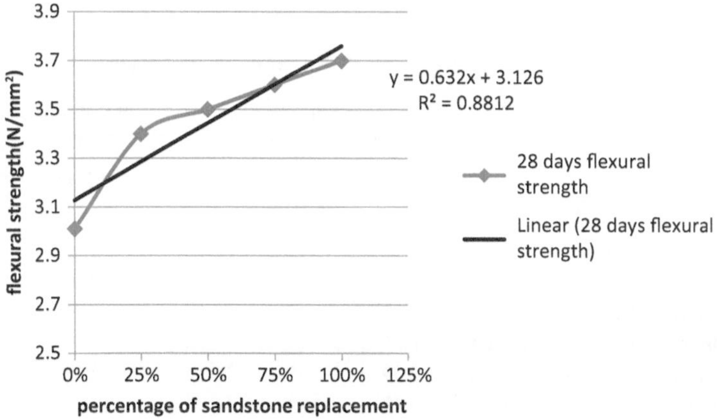

FIGURE 3.23 Variation of flexural strength with sandstone replacement.

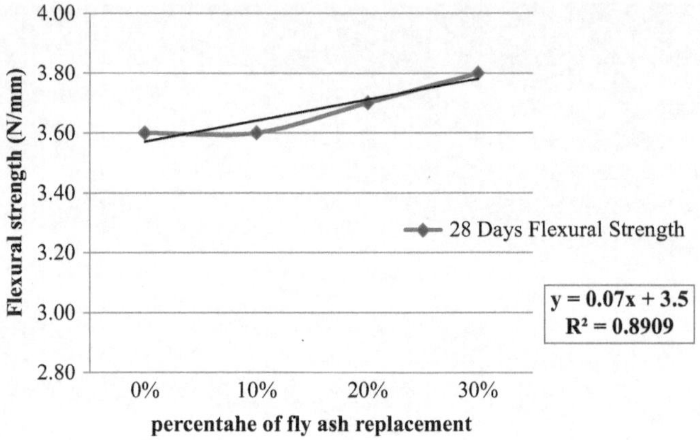

FIGURE 3.24 Variation of flexural strength with fly ash replacement.

3.4 SUMMARY

Table 3.7 shows the percentage of variation in different strengths with reference to base mix with different percentage of replacement of sand with sandstone. From the results, it can be observed that only the 7-days strength of 30% fly ash replacement along with sandstone acquired lower strength compared with base mix. Apart from that lone mix, all others showed an increasing trend. Flexural strength and split tensile strength exhibited higher variation when compared with compressive strength. So, it can be stated that sandstone and fly ash can be effectively used as replacement for sand (fine aggregates) and cement, respectively, without compromising on any required properties of concrete.

TABLE 3.7
Percentage of Variation in Strengths with Respect to Base Mix

	Compressive Strength			Split Tensile Strength	Flexural Strength
Mix	3 Days	7 Days	28 Days	28 Days	28 Days
25% sandstone replacement	13.88	17.21	23.34	3.25	3.3
50% sandstone replacement	15.12	18.12	24.34	3.39	3.4
75% sandstone replacement	15.79	18.32	24.38	3.46	3.5
100% sandstone replacement	15.77	18.57	24.79	3.46	3.6
100% sandstone replacement + 10% FA	8.85	5.81	10.30	11.17	20.00
100% sandstone replacement + 20% FA	3.99	3.89	10.16	12.29	23.33
100% sandstone replacement + 30% FA	3.19	−3.83	11.93	17.85	26.67

REFERENCES

Charles, B., Zhua, J., Jensena, W. and Tadrosb, M. 2005. High-percentage replacement of cement with fly ash for reinforced. *Cement and Concrete Research*, 35, 1088–1091.

Dhir, R.K., Munday, J.G.L. and Ong, L.T. 1986. Investigation on engineering properties of opc/pulverized fly ash concrete-deformation properties. *The Structural Engineer*, 64B(2), 26–42.

Duxson-Loya, E.I., Allouche, E.N. and Vaidya, S. 2007. Mechanical properties of fly ash based geopolymer concrete. *ACI Material Journal*, 108, 300–306.

Fernandez-Jimenez, A. and Palomo, A. 2003. Characterisation of fly ash: Potential reactivity as alkaline cements. *Fuel*, 82(18), 2259–2265.

Fontanini, P.S.P., Pimentel, L.L., Jacintho, A.E. and Migliato C. L. 2013. Brazilian geological sandstone characterization and its utilization as aggregate in structural concrete. *Applied Mechanics and Materials*, 271–272, 141–146.

Kesavulu, N.C. 2009. *Textbook of Engineering Geology*. MacMillan, New Delhi.

Kiliç, A., Atis, C.D., Teymen, A., Karahan, O., Zcan, F.O., Bilim, C. and Zdemir, M.O. 2007. The influence of aggregate type on the strength and abrasion resistance of high strength concrete. *Cement & Concrete Composites*, 30, 290–296.

Kumar, P.S., Mannan, M.A., Kurian, V.J. and Achyutha, H. 2005. Effect of crushed sandstone sand on the properties of high performance concrete. *Journal of Civil Engineering Research and Practice*, 2(2), 1–11.

Kumar, P.S., Mannan, M.A., Kurian, V.J. and Achyutha, H. 2007. Investigation on the flexural behaviour of high-performance reinforced concrete beams using sandstone aggregates. *Building and Environment*, 42, 2622–2629.

Maharana, J.K. and Patel, A.K. 2013. Physico-chemical characterization and mine soil genesis in age series coal mine overburden spoil in chromosequence in a dry tropical environment. *Phylogenetics & Evolutionary Biology*, 1(1), 1–7.

Mucteba, U. 2012. The influence of coarse aggregate type on mechanical properties of fly ash additive self-compacting concrete. *Construction and Building Materials* 37, 533–540.

Palomo, A., Grutzeck, M.W. and Blanco, M.T. 1999. Alkali-activated fly ashes: A cement for the future. *Cement and Concrete Research*, 29(8), 1323–1329.

Prashant, M., Ghosh, C.N. and Mandal, P.K. (2010). Use of crushed and washed overburden for stowing in underground mines: A case study. *Journal of Mine Metals and Fuels*, 58(1&2), 7–12.

Punmia, B.C. and Jain, A.K. 2005. *Soil Mechanics and Foundations*. Laxmi Publications, New Delhi, 98–101.

Rai, A.K., Paul, B and Singh, G. 2010. A study on the bulk density and its effect on the growth of selected grasses. *International Journal of Environmental Sciences*, 1(4), 677–684.

Santos, Cassiano R.D., Juarez, R.D., Amaral, F., Rejane, M.C.T. and Ivo, A.H.S. 2013. Use of coal waste as fine aggregates in concrete. *Geomaterials Scientific Research*, 3(2), 54–59.

Sayar, Y., Pal, A., Singh, S. and Yousuf, D. 2012. A study of physico-chemical characteristics of overburden. *Global Journal of Science Frontier Research*, 12(1), 7–13.

Shetty. 2005. *Concrete Technology*. S. Chand & Company Ltd., New Delhi.

Twardowska, I., Allen, H.E., Kettrup, A.F. and Lacy, W.J. 2004. *Solid Waste: Assessment, Monitoring and Remediation: Volume 4 (Waste Management)*. Pergamon. ISBN: 9780080541471.

4 Utilization of Laterite Waste in Concrete

Manjunath B. and Ram Chandar Karra

CONTENTS

4.1 INTRODUCTION

Francis Buchanan-Hamilton in 1807 first described and named a laterite formation in coastal Kerala and Karnataka states of southern India. He named it laterite from the Latin word later, which means a brick. Laterite rock can be cut into bricks and used in construction. Laterites are the products of intensive and long-lasting tropical rock weathering which is intensified by high rainfall and elevated temperatures.

DOI: 10.1201/9781003268499-4

These soils are rich in iron and aluminium and are rusty red in colour due to the presence of iron oxides. These laterites occur as extensive cappings in the Western Ghats and in coastal plains. Their thickness ranges from a few centimetres to as much as 60 m. It was estimated that laterites cover about one-third of the Earth's continental land area (Tardy 1997).

Laterite and lateritic soils are widely spread in India and cover an area of 2.48 lakh sq km. They also occur at lower levels and in valleys in several other parts of the country. Laterite and lateritic soils have a unique distinction of providing valuable building material. When moist, laterites can easily be cut with a spade into regular-sized blocks. The art of quarrying laterite material into masonry is suspected to have been introduced from the Indian subcontinent (Mohita 2015).

Approximately 30% of the laterite is wasted as scrap while cutting. So, the laterite scrap produced from the laterite quarry due to cutting needs to be processed and reused so that the finer particles of this scrap doesn't get washed away by the rains in the coastal regions, thereby turning the colour of the river body or water body into reddish brown colour causing environmental pollution. The colour of laterite is due to iron oxide and aluminium oxide present in the laterite.

On the other hand, concrete is one of the commonly used construction materials, but there is a need to develop new and sustainable technologies to use waste produced from laterite quarries in the manufacture of concrete blocks.

The strength of the laterite concrete is mainly dependent on the aggregate-cement bond whereas the physical properties of the aggregates are only of secondary importance. Concretionary laterite gravels are potential alternative cheap sources of aggregates for structural concrete (Akpokodje and Hudec 1992).

The laterized concrete has a higher rate of moisture loss to the environment than normal concrete. Similarly, the rate of moisture absorption of laterized concrete is higher than that for normal concrete of the same mix proportion and water–cement ratio (Salau and Balogun 1998).

4.1.1 Compressive Strength

The compressive strength of laterite aggregate concrete is considerably lower than that of gravel or crushed granite aggregate concrete (Krishna Raju and Ramakrishnan 1972).The compressive strength increased with age up to 40% replacement; later, it decreased with an increase in the percentage replacement level of sand. The strengths of laterite concrete generally increases with age but decreases with increase in the replacement level of sand by laterite (Udoeyo et al. 2005). Compressive strength of sandcrete block decreases at a decreasing rate as percentage of laterite content increases (Ata et al. 2007). Unconfined compressive strength decreased with higher fines content. The compressive strength of the concrete samples increased as the days of hydration increased (Ogunbode and Olawuyi 2008).

Compressive strength of laterite-cement mix increased with increase in percentage of cement content up to 20% but decreased at cement content above 20%. The maximum strength of laterite-cement mix was achieved at moisture content slightly greater than or equal to optimum moisture content for the particular mix (Aguwa, 2009). Increase in water–cement ratio causes reduction of compressive strength of both concrete and lateritic concrete mixes. However, the compressive strength of both concrete and lateritic concrete mixes increases with age (Alawode and Idowu 2011).The strength of laterite cubes (whether plain or reinforced with kernel shells) is better compared with the strength of sandcrete blocks popularly used in day-to-day building construction work as partitions (Adebayo 2012).

Presence of coarse-grained good-quality laterite in the making of concrete would not only maintain the ultimate strength of the concrete but could also improve some of its mechanical properties. For optimum performance of laterized concrete as structural members of a building, the content of laterite replacement in the concrete should not exceed about 25% (one quarter of sand) in a standard mix (Olutoge et al. 2014). It was found that 0.55 water/cement ratio produced higher compressive strength, tensile strength and better workability for M20 mix proportion. These results are better compared with those of conventional concrete. The concrete was found to be suitable for use as structural members for buildings and related structures, where laterite content did not exceed 50% (Jayaraman et al. 2014).

The cube compressive strength of laterite concrete decreases with an increase in temperature, which is in line with the findings of Ikponmwosa and Salau (2010) and Udoeyo et al. (2010). Compressive strength of laterite concrete decreased in a similar manner to that of plain concrete when subjected to elevated temperatures between 200 and 600°C (Udoeyo et al. 2010). Replacement of fine aggregate by laterite had a reasonable compressive strength for temperature applications up to 100°C whereas plain concrete and other percentage replacement suffered a reduction in compressive strength as the temperature increased. The laterized concrete should not be exposed to constant rainy and dry season as it could result in reduction of its compressive strength (Ige, 2013).

Compressive strength of both the concrete and laterized concrete specimens decreased with increasing acid concentration, immersion period and laterite content (Olawuyi et al. 2012).The compressive strength of laterite/sand block decreased with increasing magnesium sulphate concentration and the exposure period. Laterite/sand block made with ordinary Portland cement cannot be recommended for use in sulphate-laden environment (Olugbenga et al. 2014). Concentration levels of magnesium sulphate solution influenced the severity of its attack on conventional and laterized concrete. The higher the magnesium sulphate solution concentration, the higher the loss in compressive strength of concrete exposed to it (Olubunmi et al. 2014).

4.1.2 GGBS

There was an appreciable increase in the workability of concrete with increasing percentage replacement of cement with GGBS (Ground-Granulated Blast-furnace Slag); therefore, water–cement ratio can be reduced keeping the slump constant, which will result in an increase in compressive strength. Even if water–cement ratio is decreased using water reducers, compressive strength can be increased up to strength of normal cement concrete (Khan et al. 2014). Strength of concrete with different GGBS content is according to substitute content and water–cement ratio. When the substitute content is higher, the early strength is lower, but the 28 and 60 day strength is almost similar (Ling et al. 2004). The compressive strength of GGBS concrete increases as the GGBS content is increased up to an optimum point, after which the compressive strength decreases. There is an optimum level for the efficient use of GGBS content, which yields the highest strength (Oner et al. 2007). After an optimum point, at around 9% of GGBS and 40% of Fly ash of the total binder content, the further addition of GGBS and fly ash does not improve the compressive strength, split tensile strength and flexural strength. The strength development of GGBS concrete was affected by the curing temperature. Low curing temperature would result in low early strength of GGBS concrete. For high temperature curing at 75°C, the 28-day strengths fell short of their design strength and there may be a need to limit the peak temperature of concrete in mass pours in practice (Leung and Wong 2011). After 3 days of curing the increase in the compressive strength is not significant (Supraja and Kanta Rao 2013).

The partial replacement of OPC in concrete by GGBS not only provides the economy in the construction but it also facilitates environmental friendly disposal of the waste slag which is generated in huge quantities from the steel industries. The increase in percentage of GGBS results in decrease in strength of concrete (Patil et al. 2013). The flexural strength increases with the increase in percentage of GGBS and Robo sand. The optimum percentage of GGBS replacing cement is 50% for getting maximum compressive strength. The test results proved that the compressive strength, split tensile strength and flexural strength of concrete mixtures containing GGBS and Fly ash increase as the amount of GGBS and Fly ash increase.

GGBS can be used as alternative material for the cement. The compressive strength, splitting tensile strength, flexural strength and modulus of elasticity increases with increasing GGBS content. The drying shrinkage shows a slight increment with GGBS (Rughooputh and Rana 2014). Slag replacement by weight decreases the strength of Concretes in short term when compared to control Portland cement concrete. However, in long term, concrete containing slag exhibits an equivalent or a greater final strength than that of control normal Portland cement concrete (Ramlekshmi et al. 2014). When the replacement percentage is increased from 0 to 65% strength is observed to increase up to optimized value. The optimum percentage is observed at the replacement level of 40% (Guruvu et al. 2014).

4.1.3 Workability

Workability is often referred to as the ease with which a concrete can be transported, placed and consolidated without excessive bleeding or segregation. Workability is the ability of a fresh concrete mix to fill the form/mould properly with the desired work (vibration) and without reducing the concrete's quality. Workability depends on water content, aggregate (shape and size distribution), cementitious content and age (level of hydration) and can be modified by adding chemical admixtures like super plasticizer. Raising the water content or adding chemical admixtures increases concrete workability. Excessive water leads to increased bleeding (surface water) and/or segregation of aggregates (when the cement and aggregates start to separate), with the resulting concrete having reduced quality.

The laterized concrete has a higher rate of moisture loss to the environment than normal concrete, similarly; the rate of moisture absorption of laterized concrete is higher than that of normal concrete of the same mix proportion and water–cement ratio (Salau and Balogun 1998). The workability of laterite concrete increases with higher replacement level of sand by laterite, while the percentage of water absorption of the concrete decreases with increase in replacement level (Udoeyo et al. 2005). The workability of the concrete specimens decreased as the content of laterite and volcanic ash increased (Ogunbode et al. 2008). The workability of laterite concrete increases while the percentage of water absorption by the concrete decreases with increase in replacement level of sand by laterite. The workability of concrete using lateritic sand and quarry dust as fine aggregates was found to have the same trend with normal concrete (Ukpata et al. 2012).

GGBS requires about 3% lesser water content in comparison to OPC for the equal slump requirement. This is due to the smooth surface texture of the slag particles that delay the chemical reaction and increase the setting time. As the proportion of GGBS is increased, the heat of hydration is reduced. This is beneficial in large concrete pouring that enable reduced temperature which will reduce the probability of thermal cracking. Since GGBS slowly reacts with water as compared to Portland cement, therefore stiffening/setting time of concrete is high. The setting time will be greater at high replacement levels above 50% and at lower temperatures (below 10°C) (Newman 2003). With considerable sustainability benefits, GGBS also lowers early-age temperature rise, reducing the risk of thermal cracking in large pours. Compared to hot air oven curing and curing by direct sunlight, oven-cured specimens give the higher compressive strength but sunlight curing is convenient for practical conditions (Supraja and Kanta Rao 2013).

The degree of workability of concrete was normal with the addition of GGBS up to 40% replacement level for M35 grade concrete (Arivalagan et al. 2014). The setting time of concrete is influenced by many factors, in particular temperature and water/cement ratio. With GGBS, the setting time will be extended slightly, perhaps by about 30 minutes. The effect will be more pronounced at high levels of GGBS and/or low temperatures (Guruvu et al. 2014). By adopting same critical mix and replacing cement by GGBS, it is found that by increasing the percentage of GGBS; workability increases but strength decreases. In order to increase the strength, cement is replaced by combination of GGBS and RHA (Gadpalliwar et al. 2014). The workability of the concrete increases with the increase in the replacement levels. Water absorption test concluded that durability of the concrete increases with the increase in replacement levels (Garg and Khadwal 2014). Addition of slag in the mixture decreases the setting time and the presence of slag in the mix reduces the slump value, thereby increasing the degree of workability (Krishnaraja et al. 2014).The partial replacement of OPC with GGBS improves the workability but causes a decrease in the plastic density of the concrete (Rughooputh et al. 2014).

4.1.4 Tensile and Flexural Strength

The tensile strength increases with age but decreases with increase in replacement level of cement and sand. The ratio of the 7 days to 28 days strength shows that there is a reduction in the rate of strength gain as the replacement level increases. Shear strength parameters of lateritic soils are affected by leachate contamination (Goswami and Choudhury 2013).

GGBS can be used as alternative material for the cement. Based on the results, the compressive, split tensile, flexural and Pull out strengths are increased as the percentage of GGBS increased up to 40% and above decreased. Split tensile strength and modulus of rupture of 50% replacement mix concrete give better result when compared with all the mixes (Krishnaraja et al. 2014).The compressive strength, splitting tensile strength, flexural strength and modulus of elasticity increase with the increasing GGBS content. The drying shrinkage shows a slight increment with GGBS (Rughooputh et al. 2014). The flexural strength increases with the increase in percentage of GGBS and Robo sand. The optimum percentage of GGBS replacing cement is 50% for getting maximum compressive strength.

4.1.5 Physico-Chemical Properties

The shrinkage strain of laterized concrete is several times greater than that of normal concrete depending on the laterite content, the higher the laterite content the greater the shrinkage strain (Salau and Balogun 1998). Compressive strength and specific gravity of laterite blocks decrease with depth within a quarry, while there is an increase in water absorption; this may be attributed to the increase in clay content (kaolinite).

The specific gravity decreased with increased fines from 2.65 to 2.53. The Atterberg limits (plastic limit, liquid limit, plasticity index and linear shrinkage) values increased with higher fines content. The particle sizes of the reconstituted soils generally increased with increasing fines content resulting in a decrease in grading modulus. Strength and hardness of laterite were attributed to the presence of iron minerals and their compact arrangement. Dense packing of crystalline iron minerals offered resistance to failure, which may be the reason for high strength of laterite. High water absorption capacity of the material is attributed to the abundance of clay in the material matrix (Kasthurba et al. 2007).

The engineering properties and chemical characteristics of the natural lateritic soil are susceptible to change when compared with the initial characteristics of the soil. The studies clearly indicated that under soaked conditions the pH of the water has strong influence on the chemical characteristics of lateritic soil. The influence on the engineering properties of the soil is also observed. The method of sample preparation prior to testing has significant effect on the index properties of lateritic soils. Although there is no significant change in the mineralogy of the air dried and oven-dried soil, however, drying has more effect on the Atterberg limits and grain size distribution of lateritic soils. The plasticity index values and clay content of soils tested decreased due to drying. However, repeated wetting and drying may have significant effect on the structure and affect the soil properties. The chemical properties of lateritic soil such as pH, organic matter and iron content decreased after oven drying (Sunil and Krishnappa. 2012).

Both linear shrinkage and unit weight of bricks decreased with increase in the percentage of RHA content. The compressive strength of lateritic soil-clay mixed bricks increased almost linearly with increase in the percentage content of RHA (Rahman 1988).The study has proved that the influence of the parent rock factor on engineering index properties such as the specific gravity of grains, plasticity index and grain size distribution characteristics of the soils studied was significant (Adeyemi 1994).

Atterberg limits of subgrade are influenced due to fines percentage. As fines percentage is more, liquid limit and plastic limit is more and it is observed that percentage of fines has effect on void ratio for all subgrades (Chukka et al. 2012). The leachate can alter the Attterberg limit of lateritic soils. All the leachate-contaminated lateritic soil samples showed an increase in liquid limit and plasticity index values. The increase in liquid limit (WL) and plasticity index (IP) of the lateritic soil is attributed due to change in nature of pore fluid which is shown by increase in clay content of the soil (Goswami and Choudhury 2013).The use of laterite fines as a partial replacement has a significant influence on the engineering properties of bricks. Bricks with laterite fines replacing the natural sand can satisfactorily perform as a masonry unit when the laterite fines content does not exceed 30% (Emmanuel et al. 2014).

Based on the results, it was found that both additives can decrease the dry density and increase the laterite soil strength approximately fourfold in comparison with the natural soil; it was found that the treatment of laterite had a marginal impact on the thermal characteristics of the soil (Marto et al. 2014).

Partial replacement of Portland cement by GGBS generally increases the resistance of concrete to sulphate attack and chloride attack. The use of a minimum of 50% GGBS can be a suitable method of preventing alkali-silica reaction, if the alkalis are not contributed by sources other than cementitious materials. Higher replacement level of GGBS in concrete significantly reduces the chloride content in concrete. The results also show that chloride concentration decreases with increasing depth into concrete (Sumadi et al. 1999). GGBS can be used as one of the alternative materials for the cement. From the experimental results, it was found that 50% of cement can be replaced with GGBS (Malagavelli et al. 2010). The inclusion of GGBS would improve the concrete's ability to resist chloride penetration, but the GGBS replacement percentage will need to be at least 70% for this purpose. The source of GGBS does not appear to have a significant effect on the performance of GGBS concrete as long as the GGBS complies with the relevant standards (Leung et al. 2011).

Use of GGBS accelerates the hydration of ordinary Portland cement at early hours of hydration. Consistency of cement decreased with the increase in GGBS content (Siddique and Kaur 2012). In respect of sulphate attack that resistance of Portland cement binders is greatly enhanced by the use of high quantities of GGBS. Test results showed that GGBS combination produced a binder that was comparable or outperformed the sulphate-resistant Portland cement concrete (Connell et al. 2012). GGBS replacement enhances lower heat of hydration, higher durability and higher resistance to sulphate and chloride attack when compared with normal ordinary concrete. On the other hand, it also contributes to environmental protection because it minimizes the use of cement during the production of concrete (Parthiban et al. 2013).

4.1.6 California Bearing Ratio Test

Lateritic aggregates are good materials for road chippings and concrete aggregates although they give results slightly inferior to those obtained from igneous aggregates (Madu 1980). The sedimentary laterite soil constitutes a good engineering construction material as it has already been successfully used as base and sub-base material in road construction. Laterite soil is also suitable for use as fill materials in embankment and dam construction (Olufemi 1989). In comparison with other typical construction fills available in Southeast Asia, compacted lateritic soils produce excellent shear strengths and California Bearing Ratio (CBR) values (Indraratna and Nutalaya, 1991). These lateritic clayey gravels have good workability as engineering construction materials and are rated fair to good as road aggregates in terms of probable *in-situ* behaviour based on water absorption values. Relationships between maximum dry unit weight and optimum moisture content as well as those between CBR and some derived soil parameters such as grading modulus, plasticity product or plasticity modulus can be described using a third-order polynomial function. The soils are mainly clayey gravels which have good workability as engineering construction materials and rated as good for gravel road surfacing applications (Nwaiwu et al. 2006).

CBR (both soaked and Unsoaked) decreased with increasing fines content, while the UCS, however, increased with increasing fines content to about 60% before it decreased rapidly to zero. As the fines content increased, the compaction characteristics of the soil samples reduced, with Optimum Moisture Content (OMC) increasing and the Maximum Dry Density (MDD) decreasing (Ayodele et al. 2009).The strength increases proportionally with increasing cement content. Stabilized lateritic soil can be used as road base course. Only about 3% by weight of the Portland cement is enough to stabilize lateritic soil to meet the Department of Highways specification and can be an economical substitution for crushed rock. Furthermore, the use of stabilized lateritic soil decreases environmental problems in decreasing demand on crushed rock. Unnconfined compressive strength as well as the CBR of cement and lime stabilized soil increased with increasing cement and lime content (Ochepo et al. 2013).

4.2 EXPERIMENTAL INVESTIGATIONS

To assess the suitability of laterite waste from quarries for partial replacement of fine aggregates for use in concrete, laboratory investigations were carried out in two stages. In the first stage, studies were carried out to understand the physical, mechanical and chemical properties of laterite soil collected from quarries. In the second stage, experiments were carried out on moulded samples to find the variation of strength with respect to the curing period and percentage increase in laterite content in fine aggregate with fixed water–cement ratio and 40% of GGBS content and 60% cement content on moulded samples.

4.2.1 Sample Collection

The primary objective while collecting a soil sample for laboratory analysis is that its composition should be representative of the conditions that exist in the field. The general procedure involves the random collection of several individual soil cores over the designated area and combining them to form a composite sample for analysis. If soil samples are carefully collected and processed, the test results will be very useful. In order to carry out the experiments, samples were collected from quarries from 1 m below the ground level. All the soil samples were collected from open trial

FIGURE 4.1 View of the quarry.

pits of depth of about 1.5 m to 2 m from natural ground level. A view of the quarry is shown in Figure 4.1. Six to eight bags of 35 to 40 kgs capacity of disturbed soil samples were collected from the quarry. Lateritic soil samples were air dried and thoroughly mixed to ensure homogeneity.

4.2.2 Materials Used

In general, concrete is a mixture of cement, fine aggregate, coarse aggregate and water. In order to utilize waste produced from different sources in concrete, laterite waste was used in place of fine aggregate, and GGBS from steel and iron industry waste was used to replace cement partially. The details of the respective materials used for the specimen preparation are discussed in the sections below.

Cement: Cement is a binder substance that sets and hardens and can bind other materials together. Cement used in construction can be characterized as being either hydraulic or non-hydraulic in nature, depending upon the

TABLE 4.1
Physical Properties of Cement

Properties	Experimental Results	Standards
Specific Gravity	3.12	3.10 – 3.15
Initial setting time (min)	68	30 (min)
Final setting time (min)	360	600 (max)
Fineness (%)	1.75	10 (max)

ability of the cement to be used in the presence of water. The most important uses of cement is as a component in the production of mortar in masonry and in concrete as combination of cement and as an aggregate to form a strong building material. Ordinary Portland cement is the most common type of cement generally used around the world. Locally available 43 grade cement of ACC brand is used for experimental purpose. The physical properties of cement are given in Table 4.1.

Aggregates: Construction aggregate, or simply aggregate, is a broad category of coarse particulate material used in construction, including sand, gravel, crushed stone, slag, recycled concrete and geosynthetic aggregates. Aggregates are the most mined materials in the world. The aggregate serves as reinforcement to add strength to the overall composite material. These aggregates are mainly classified into two types, namely, fine aggregate of size less than 4.75 mm and coarse aggregate of size more than 4.75 mm. Locally available river sand is used as fine aggregate. Gravels constitute major part of coarse aggregates. The sample of coarse aggregates in a pan is shown in Figure 4.2. The physical properties of aggregates are given in Table 4.2.

Laterite soil: Laterite is well known in Asian countries as a building material for more than 1000 years (Schellmann et al. 1994). Laterites are the products of intensive and long-lasting tropical rock weathering which is intensified by high rainfall and elevated temperatures. These soils are rich in iron and aluminium and are rusty red in colour due to the presence of iron oxides (Schellmann et al. 1994). Figure 4.3 shows sample of laterite soil in a pan.

Ground-Granulated Blast-furnace Slag (GGBS): GGBS is waste product obtained from Iron ore industries. GGBS is obtained by quenching molten iron slag (a by-product of iron and steel-making) from blast furnace in water or steam to produce a glassy granular product that is then dried and ground into a fine powder (Siddique and Kaur 2012). This powder is added to concrete to improve the strength properties of concrete and hence called admixture. The specific gravity of GGBS used for the experimental purpose is 2.9.

4.2.3 Mix Design

Mix ratio of M20, i.e. 1:1.5:3, is considered for design of both conventional concrete and pozzolanic concrete. The cement portion is made up by 40% GGBS and

FIGURE 4.2 Coarse aggregates.

TABLE 4.2
Physical Properties of Aggregates

Property	Fine Aggregates	Coarse Aggregates
Specific Gravity	2.62	2.72
Water Absorption	0.48%	0.48%
Moisture Content	Nil	Nil
Maximum size (mm)	4.23	19.5

60% Cement. The fine aggregate portion is a combination of both river sand and laterite in proportions of 100%–0%, 25%–75%, 50%–50% and 75%–25%. Water–cement ratio of 0.5 is adopted. All the materials are mixed in order to produce concrete. Produced concrete is placed in cube specimens of size 150 mm × 150 mm × 150 mm for compressive strength, cylindrical specimens of 150 mm diameter and 300 mm height for split tensile strength and beams of size 500 mm × 100 mm × 100 mm for flexural strength. All the specimens were kept in open air for 24 hours after casting and demoulded and placed in water for curing. Overall, a total of 60 cubes and 20 prisms and 20 cylinders are tested. 3, 7 and 28 days testing is carried out for cubes and 28 days for both cylinders and prisms.

FIGURE 4.3 Laterite soil in a pan.

TABLE 4.3
Mass of Ingredients of Concrete for Percentage Replacements

		Percentage Replacements			
Ingredients	**Control Mix**	**0%**	**25%**	**50%**	**75%**
Cement (Kgs)	34.52	20.71	20.71	20.71	20.71
GGBS (Kgs)	–	12.79	12.79	12.79	12.79
Fine Aggregate (Kgs)	66.44	66.44	49.83	33.22	16.61
Laterite Soil (Kgs)	–	–	16.61	33.22	49.83
Coarse Aggregate (Kgs)	99.23	99.23	99.23	99.23	99.23
Water (Kgs)	16.90	16.90	18.10	20.52	22.32

Table 4.3 contains weights of concrete ingredients required for 12 cubes, 4 cylinders and 4 beams for each of the combination, i.e. a total of 60 cubes, 20 cylinders and 20 beams have been casted for the experimental procedure. The total numbers of specimens casted for the investigation are mentioned in Table 4.4.

4.3 PROPERTIES OF LATERITE

The experiments to find out the physical properties of laterite soil, physico-mechanical and chemical properties were carried out conforming to particular IS codes.

TABLE 4.4
Total Number of Specimens Casted for the Investigation

		Total Number of Elements Cast				
		Cubes			Cylinders	Beams/Prisms
Sl no.	Percentage Replacements	3 days	7 days	28 days	28 days	28 days
1	Control mix	4	4	4	4	4
2	0	4	4	4	4	4
3	25	4	4	4	4	4
4	50	4	4	4	4	4
5	75	4	4	4	4	4

4.3.1 Grain Size Distribution

Laterite soil in hydrometer analysis in a cylindrical jar is shown in Figure 4.4. Table 4.5 gives the grain size distribution of laterite soil. Figure 4.5 shows the grain size distribution curve of laterite soil. The classification of soil was done based on percentage of fraction retained on each sieve. The particles retained on 4.75 mm sieve are

FIGURE 4.4 Laterite soil in measuring cylinder for hydrometer analysis.

TABLE 4.5
Grain size distribution for laterite soil

Particle	Percentage
Gravel	17.6
Sand	47.3
Silt	29.8
Clay	5.3

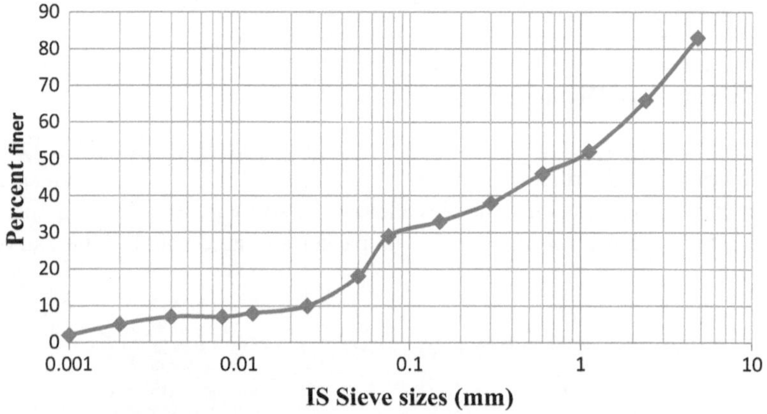

FIGURE 4.5 Grain size distribution curve of laterite soil.

gravels. The particles retained on 600 μm are called sand. The particles passing through 75 μm are called silt. The particles retained on 2 μm are called clay.

4.3.2 Specific Gravity and Water Absorption

A total of four trials were carried out on laterite specimen for specific gravity and water absorption, calculations and the results are tabulated in Table 4.6. A coarse aggregate in pycnometer is shown in Figure 4.6. Specific gravity of laterite soil (2.54) is close to the calculated value for river sand (2.64), and this can be used based on specific gravity value. The water absorption value for laterite soil is too high, i.e. 10.86%, which is approximately 22 times of water absorption of river sand which is 0.5%. This has given way for addition of more water during mixing of concrete.

4.3.3 Atterberg Limits

A total of four representative samples for liquid limit and three representative samples for plastic limit were carried out as per IS: 2720 1985 part 5. Results of the tests are tabulated in Table 4.7. The flow curve graph that requires calculating liquid limit

TABLE 4.6
Specific Gravity and Water Absorption Results

	Trial 1	Trial 2	Trial 3	Trial 4
Weight of empty pycnometer with closing lid (W1) gms	662	662	662	662
Weight of pycnometer + 1/3rd of sample (W2) gms	1054	1142	947.6	1094
Weight of pycnometer + sample + Distilled water (W3) gms	1781	1836	1717.2	1805.8
Weight of pycnometer + distilled water (W4) gms	1543.8	1544	1544	1547
Weight of oven-dried sample (W5)	354.7	429.7	259.8	388.1

FIGURE 4.6 Coarse aggregate for specific gravity test in pycnometer.

is shown in Figure 4.7. A figure of laterite soil in Casagrande's cup for liquid limit test is shown in Figure 4.8.

The liquid limit value of the sample was 42%. The plastic limit of the laterite sample was 30%. The plasticity index of the soil sample is 12%; hence, it is slightly plastic from the classification of plasticity index. From IS 1498-1970 for liquid limit of 42% and plasticity index of 12%, the free swell of soil is 50%–100% and degree of expansion is medium. From IS 1498-1970 and the results of liquid limit and plasticity index, it can be concluded that the soil is classified as SM (silty sand).

4.3.4 Slake Durability Index

A total of four trials of same sample were conducted on the oven-dried laterite rock samples. A figure of Slake durability test under progress is shown in Figure 4.9.

TABLE 4.7
Atterberg Limits Results

Description	Liquid Limit Test				Plastic Limit Test		
Determination No.	**1**	**2**	**3**	**4**	**5**	**6**	**7**
No. of drops	42	36	30	18	–	–	–
Weight of container (gms)	28.53	32.65	29.10	29.8	34.36	29.34	28.15
Weight of container + wet soil (gms)	47.92	52.38	43.61	51.46	47.13	39.87	38.24
Weight of container + dry soil (gms)	42.29	46.68	39.36	44.74	44.14	37.46	35.92
Weight of water (gms)	5.63	5.7	4.25	6.72	2.99	2.41	2.32
Weight of dry soil (gms)	13.76	14.03	10.26	14.94	9.78	8.12	7.77
Water content (%)	40.92	40.63	41.42	44.98	30.57	29.68	29.84
LL =	By graph		PL =		30		

Results summary

Liquid limit (%)	Plastic Limit (%)	Plasticity index (%)
42	30	12

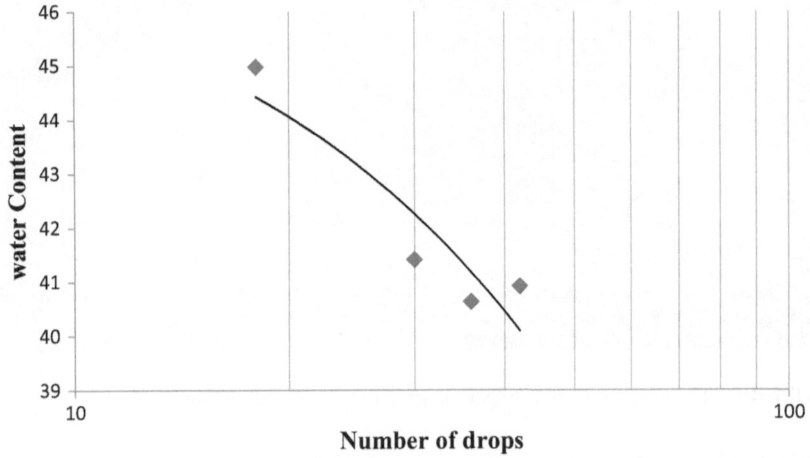

FIGURE 4.7 Liquid limit flow curve.

Table 4.8 gives results of slake durability index. As per IS 10050.1981, the laterite soil specimen sample has very high resistance to deterioration and weakening based on index value.

4.3.5 California Bearing Ratio Test

A total of three trials were carried on soaked laterite soil samples. The observations and results are given in Table 4.9, Table 4.10 and Table 4.11. A figure of CBR testing equipment is shown in Figure 4.10. The CBR value of the laterite soil under 4 days

FIGURE 4.8 Laterite sample in Casagrande's apparatus.

FIGURE 4.9 Slake durability test under progress with laterite sample.

TABLE 4.8
Slake Durability Index Test Results

Trials	1	2	3	4
Drum weight + sample 'A'	2140	2178	2120	2195
Drum weight + oven-dried sample 'B'	2130	2167	2112	2183
Drum weight + oven-dried sample 1st cycle 'C'	2102.2	2142.8	2078.7	2159.3
Drum weight + oven-dried sample 2st cycle 'D'	2093.2	2132.4	2070.1	2151.6
Empty weight of the Drum 'E'	1661	1662	1660	1662
Slake durability index (%)	90.22	91.16	89.13	91.85
Average slake durability index		90.59		

TABLE 4.9
Details of Laterite Soil for CBR Test

Trials	1	2	3
Soil sample passing through		20 mm IS sieve	
Dry density of soil, g_d (gm/cc)	1.83	1.92	1.86
Optimum moisture content (OMC) (%)	14.30	13.12	13.90
Diameter of specimen, d (cm)	15.00	15.00	15.00
Height of specimen, h (cm)	12.50	12.50	12.50
Volume of the CBR mould V (cm^3) ($p*d^2/4$)*h	2209.82	2209.82	2209.82
Weight of the soil taken for test (gms)	4043.97	4218.26	4135.52
Volume of water to be added (ml)	578.29	589.70	580.98

TABLE 4.10
Observations of CBR Test

Penetration (mm)	Dial Gauge Reading (Divisions) Load			Load in KN			Load in Kgs		
	Trials			Trials			Trials		
	1	2	3	1	2	3	1	2	3
0.00	0	0	0	0.00	0.00	0.00	0.00	0.00	0.00
0.50	12	10	9	0.72	0.60	0.54	73.51	61.26	55.13
1.00	19	20	16	1.14	1.20	0.96	116.39	122.52	98.02
1.50	39	42	39	2.34	2.52	2.34	238.91	257.30	238.91
2.00	56	55	50	3.36	3.30	3.00	343.06	336.94	306.3
2.50	66	60	69	3.96	4.02	3.96	404.32	410.44	404.32
3.00	70	74	71	4.20	4.44	4.26	428.82	453.32	434.95
4.00	85	82	80	5.10	4.92	4.80	520.71	502.35	490.08
5.00	94	96	95	5.64	5.82	5.70	575.84	594.22	581.97
7.50	127	122	118	7.62	7.32	7.08	778.00	747.37	722.87
10.00	153	149	147	9.18	8.94	8.82	937.28	912.77	900.52
12.50	171	167	168	10.26	10.02	10.08	1047.55	1023.04	1029.17

TABLE 4.11
Results of CBR Test

	Trial 1	Trial 2	Trial 3
CBR $_{2.5\,mm}$	29.51	26.83	30.85
CBR $_{5\,mm}$	28.02	28.62	27.43
CBR value	29.51	28.62	30.85
Average CBR		29.66	

FIGURE 4.10 CBR testing machine.

soaked condition is 29.66, and it is in the range of 20–50 that is suitable for base and subgrade as per IS 1498-1970 classification.

4.3.6 Scanning Electron Microscope Analysis

Experimental setup is shown in Figure 4.11. The results of Scanning Electron Microscope (SEM) analysis of a typical sample are given in Table 4.12. The graphical view of elemental percentage of one laterite soil samples is given in Figure 4.12. The view of specimens under SEM equipment is shown in Figure 4.13.

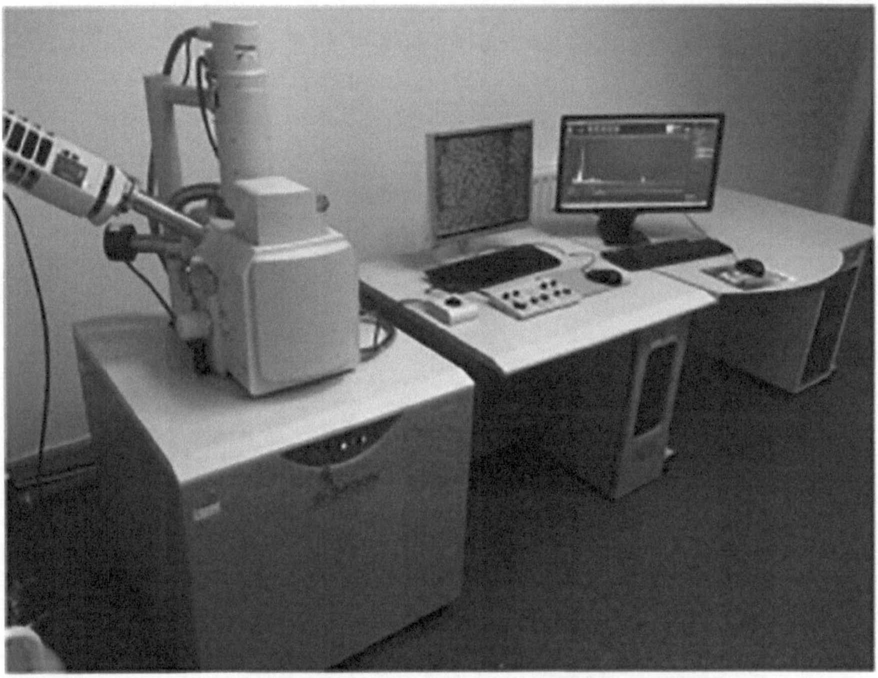

FIGURE 4.11 Scanning Electron Microscope (SEM) analysis equipment.

TABLE 4.12
SEM Analysis Results (Sample 1(view 001))

Element	(keV)	Mass%	Error%	At%	K
O K	0.525	41.29	0.08	61.43	51.0671
Al K	1.486	17.86	0.06	15.75	11.7083
Si K	1.739	12.72	0.07	10.78	8.5129
Ti K	4.508	0.76	0.09	0.38	0.7771
Fe K	6.398	27.37	0.17	11.66	27.9344
Total		100		100	

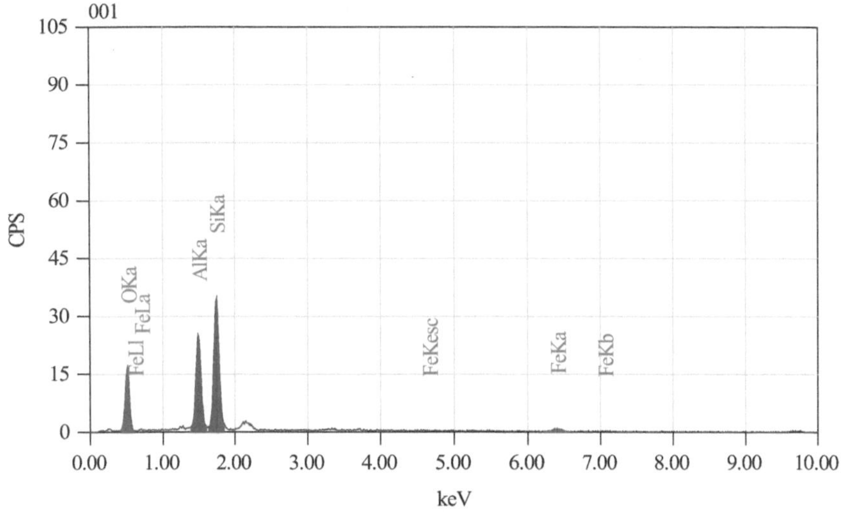

FIGURE 4.12 Graphical view of elemental % in SEM analysis sample 1(view 001).

FIGURE 4.13 Specimen view under Scanning Electron Microscope sample 1(view 001).

4.4 MECHANICAL PROPERTIES OF CONCRETE

4.4.1 Workability

A total of five trials in Slump cone were carried out for each mix at the time of casting. There was an increment in slump value for GGBS-blended cement for 0% replacement when compared to control mix, but later there was a gradual reduction in slump value as percentage of laterite content increased. The variation of slump values with percentage replacement of laterite soil with river sand is shown in Figure 4.14. This reduction in slump value was due to water absorption of laterite soil which intern affected the placing of concrete as early hardening and improper compaction has taken place. Thereby creating voids and causing reduction in strength.

4.4.2 Compressive Strength

A total of five mixes and four trials for each mix, i.e. 60 cubes, were casted. A figure of compressive strength tested on a cube is shown in Figure 4.15. Compressive strength on moulded cube samples was carried out for five different mixes including control mix at the ages of 3 days, 7 days and 28 days. There was a decrease of 6% for 25% replacement, 15% decrease for 50% replacement and 25% for 75% replacement in compressive strength when compared to 0% replacement for 3 days testing. There was a reduction of nearly 7% for 25% replacement, 13% reduction for 50% replacement and 16.5% reduction for 75% replacement in compressive strength when compared to 0% replacement for 7 days testing. There was a decrease of approximately 6.2% for 25% replacement, 11% reduction for 50% replacement and 25% reduction for 75% replacement in compressive strength when compared to 0% replacement for 28 days testing.

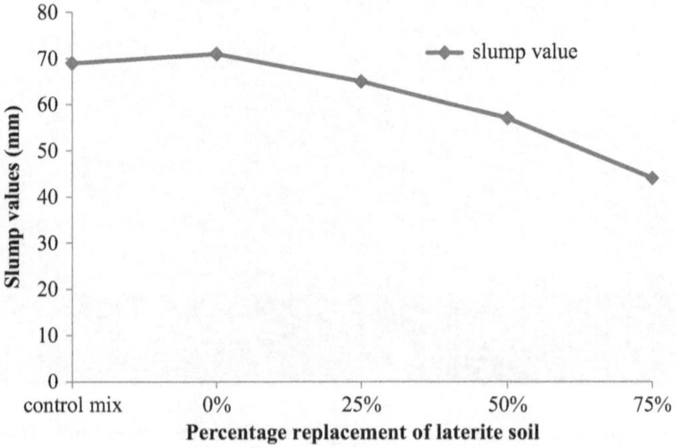

FIGURE 4.14 Variation of slump values for different mixes.

FIGURE 4.15 Cubes after compressive strength test.

This was due to the presence of laterite soil which has more water absorption capacity by which plasticity increases, which in turn left more voids in mix, thereby decreasing the workability thereby decreasing the strength. The variation of compressive strength for different mixes is given in Figure 4.16.

4.4.3 Flexural Strength

A total of four trials for each mix, i.e. 20 beams, were tested after 28 days curing. Beam after split tensile strength is shown in the Figure 4.17. There was a decrease

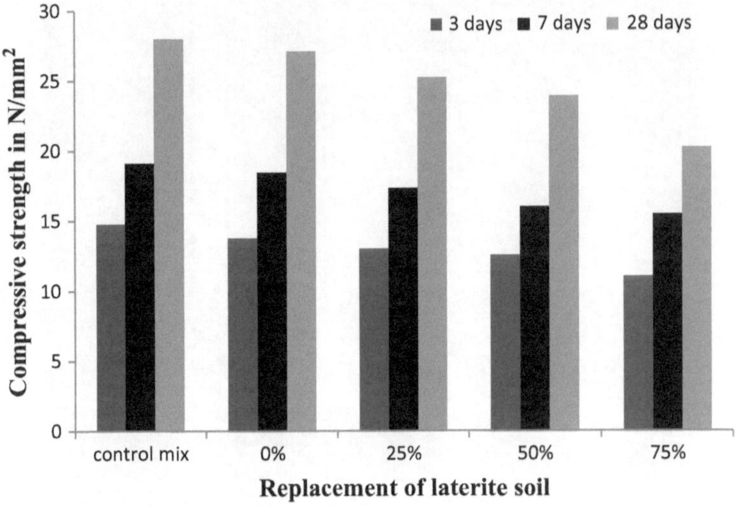

FIGURE 4.16 Variation of compressive strength for different mixes.

FIGURE 4.17 Beams after flexural strength test.

of 11.5% for 25% replacement and 28.5% for 50% replacement and 52% for 75% replacement of flexural strength when compared to 0% replacement for 28 days. The variation of flexural strength with respect to percentage replacement of sand with laterite is shown in Figure 4.18.

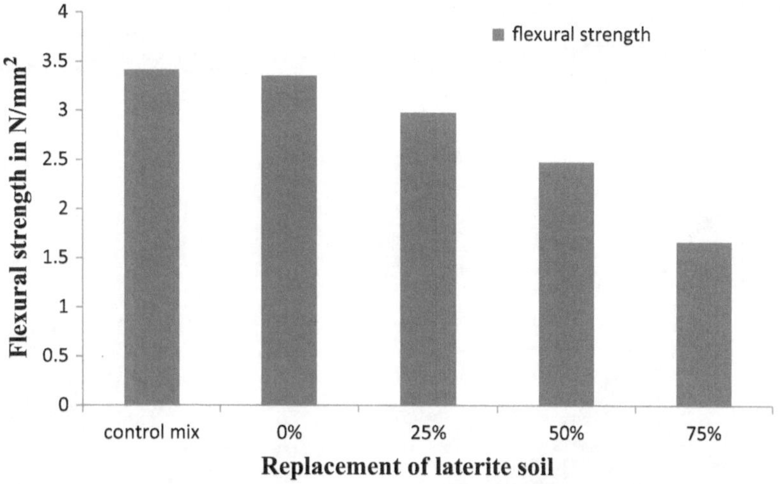

FIGURE 4.18 Variation of flexural strength for different mixes.

4.4.4 Split Tensile Strength

A total of four trials for each mix, i.e. 20 beams, were tested after 28 days curing. Cylinder in compressive testing machine for split tensile strength is shown in Figure 4.20. There was a reduction of 6.5% for 25% replacement, 15% reduction for 50% replacement and 28% reduction for 75% replacements in strength when compared to 0% replacement for 28 days cured specimens. The variation of split tensile strength with respect to percentage replacement of sand with laterite is shown in Figure 4.19.

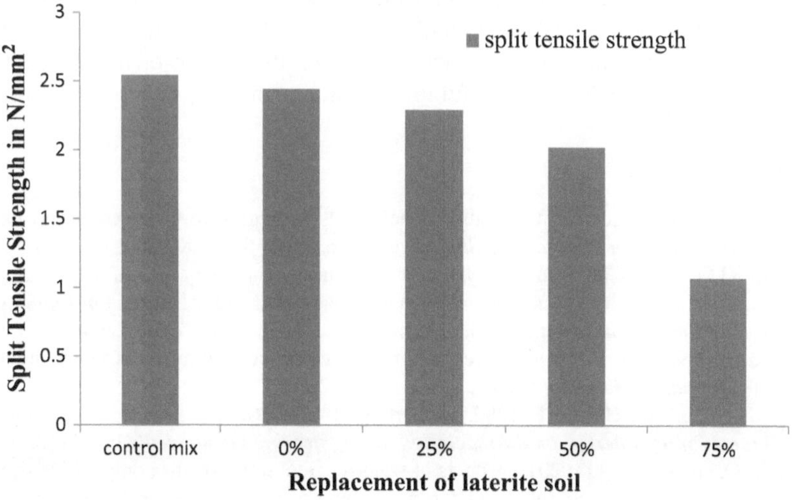

FIGURE 4.19 Variation of split tensile strength for different mixes.

FIGURE 4.20 Cylinder for split tensile strength under universal testing machine.

4.5 SUMMARY

Based on the investigations, it was found that water absorption of laterite soil is almost 10% more compared to river sand. Based on the CBR value of the soaked sample, the laterite soil specimen is found to be fit for use in subgrade and base coarse of road construction. From SEM analysis results, it is clear that there was no trace of sulphur content in the laterite soil. So the soil is safe from internal sulphate attack. After the addition of both laterite soil and GGBS, there was not much improvement in the mechanical properties of the concrete, i.e. compressive strength, split tensile strength and flexural strength, still up to 25% of laterite waste in concrete is acceptable to mix in concrete. This will reduce the burden of handling laterite waste at quarry site and will be useful to construction industry to certain extent.

REFERENCES

Adebayo, W. 2012. assessment of palm kernel shells as aggregate in concrete and laterite blocks. *Journal of Engineering Studies and Research*, 18(2), 88–93.

Adeyemi, G.O. 1994. Clay mineralogy, major element geochemistry and strength characteristics of three highway subgrade soils in southwestern Nigeria. *Bulletin of Engineering Geology and Environment*, 50(1), 5–8. http://dx.doi.org/10.1007/BF02594951.

Aguwa, J.I. 2009. Study of compressive strengths of laterite-cement mixes as a building material. *Journal of Matrerial Science*, 13(2), 114–120.

Akpokodje, E.G. and Hudec, P. 1992. Properties of concretionary laterite gravel concrete. *Bulletin of the International Association of Engineering Geology*, 46(1), 45–50.

Alawode, O. and Idowu, O.I. 2011. Effects of water-cement ratios on the compressive strength and workability of concrete and lateritic concrete mixes. *The Pacific Journal of Science and Technology*, 12(2), 99–105.

Arivalagan, S. 2014. Sustainable studies on concrete with GGBS as a replacement material in cement. *Jordan Journal of Civil Engineering*, 8(3), 263–270.

Ata, O., Olusola, K., Omojola, O. and Olanipekun, A. 2007. A study of compressive strength characteristics of laterite/sand hollow blocks. *Civil Engineering Dimension*, 9(1), 15–18.

Ayodele, A.L., Falade, F.A. and Ogedengbe, M.O. 2009. Effect of fines content on some engineering properties of lateritic soil in Ile-Ife Nigeria. *Journal of Engineering Research*, 9(32), 10–20.

Chukka, D. and Chakravarthi, V.K. 2012. Evaluation of properties of soil subgrade using dynamic cone penetration index – a case study. *International Journal of Engineering Research and Development*, 4(4), 7–15.

Connell, O. McNally, M. Richardson, C. and Mark, G. 2012. Performance of concrete incorporating GGBS in aggressive wastewater environments. *Construction and Building Materials*, 27(1), 368–374.

Emmanual, A. and Allan. A. 2014. Suitability of laterite fines as a partial replacement for sand in the production of sandcrete bricks. *Geology*, 2(4), 1–11.

Gadpalliwar, S.K., Deotale, R.S. and Narde, A.R. 2014. To study the partial replacement of cement by GGBS and RHA and natural sand by quarry sand in concrete. *IOSR Journal of Mechanical and Civil Engineering*, ISSN: 2278-1684.

Garg, C. and Khadwal, A. 2014. Behaviour of ground granulated blast furnace slag and limestone powder as partial cement replacement. *International Journal of Engineering and Advanced Technology (IJEAT)*, 3(6), ISSN: 2249-8958.

Goswami, D. and Choudhury, B.N. 2013. Atterberg's limit and shear strength characteristics of leachate contaminated lateritic soil. *Indian Journal of Resaerch*, 3(4), 91–93.

Guruvu, K.S. Ratnam, M.K.M.V. and Rangaraju, U. 2014. Experimental analysis to improve strength and durability of cement concrete with GGBS. *International Journal of Innovative Research in Science Engineering and Technology*, 3(12).

Ige, O.A. 2013. Performance of lateritic concrete under environmental harsh condition. *International Journal of Research in Engineering and Technology*, 2(8), 144–149.

Ikponmwosa, E.E. and Salau, A.M. 2010. Effect of heat on laterised concrete. *International Journal of Science and Technology*, 4(1), 33–42.

Indraratna, B. and Nutalaya, P. 1991. Some engineering characteristics of a compacted lateritic residual soil. *Geotechnical & Geological Engineering*, 9(2), 125–137.

Jayaraman, A. Senthilkumar, V. and Saravanan, M. 2014. Compressive and tensile strength of concrete using lateritic sand and lime stone filler as fine aggregate. *IJRET*, 03(1), 79–84.

Kasthurba, A.K., Santhanam, M. and Mathews, M. 2007. Investigation of laterite stones for building purpose from Malabar region, Kerala state, SW India – Part 1: Field studies and profile characterisation. *Construction and Building Materials*, 21, 73–82.

Khan, U.S., Nuruddin, M.F., Ayub, T. and Shafiq, N. 2014. Effects of different mineral admixtures on the properties of fresh concrete. *The Scientific World Journal*, 2014, 1–10.

Krishna Raju, N. and Ramakrishnan, R. 1972. Properties of laterite aggregate concrete. *Material at Construction*, 5(5), 307–314.

Krishnaraja, A.R., Sathishkumar, N.P., Sathish Kumar, T. and Dinesh Kumar, P. 2014. Mechanical behaviour of geopolymer concrete under ambient curing. *International Journal of Scientific Engineering and Technology*, 3(2), 130–132.

Leung, P.W.C. and Wong, H.D. 2011. Final report on durability and strength development of ground granulated blast furnace slag concrete. *The Government of the Hong Kong Special Administrative Region.*

Ling, Wang, Pei, T. and Yan. Y. 2004. Application of ground granulated blast furnace slag in high-performance concrete in China. *International Workshop on Sustainable Development and Concrete Technology*, 309–317.

Malagavelli, V. and Rao, P.N. 2010. High performance concrete with ggbs and robo sand. *International Journal of Engineering Science and Technology* 2(10), 5107–5113.

Marto, A. Latifi, N. and Eisazadeh, A. 2014. Effect of non-traditional additives on engineering and micro structural characteristics of laterite soil. *Arabian Journal for Science and Engineering*, 39(10), 6949–6958.

Mohita, N. 2015. Soil groups: 8 major soil groups available in India. http://www.yourarticlelibrary.com/soil/soil-groups-8-major-soil-groups-available-in-india/13902/ (March 19, 2015).

Newman, J.B. 2003. Advanced concrete technology. *Constituent Materials*. 1920, eBook ISBN: 9780080526560.

Nwaiwu, C.M.O., Alkali, I.B.K. and Ahmed, U.A. 2006. Properties of ironstone lateritic gravels in relation to gravel road pavement construction. *Geotechnical & Geological Engineering*, 24(2), 283–298.

Ochepo, J., Osinubi, K.J. and Sadeeq, J.A. 2013. Statistical evaluation of the effect of elapse time on the strength properties of lime-bagasse ash treated black cotton soil. *International Journal of Engineering Research & Technology* 2, 1–18.

Ogunbode, E.B. and Olawuyi, B.J. 2008. Strength characteristics of laterized concrete using lime-volcanic ash cement. *Environmental Technology & Science Journal*, 3(2), 81–87.

Olawuyi, B.J. Olusola, K.O. and Babafemi, A.J. 2012. Influence of curing age and mix composition on compressive strength of volcanic ash blended cement laterized concrete. *Journal of Civil Engineering Science and Application*, 14(2), 84–91.

Olubunmi, O.K. and Olugbenga, A. 2014. Durability of laterized concrete exposed to sulphate attack under drying-wetting cycles. *Civil and Environmental Research*, 6(3), 33–38.

Olufemi, O. 1989. Some properties of a sedimentary laterite soil as engineering construction material. *Bulletin of the International Association of Engineering Geology*, 39(1), 131–135.

Olugbenga, A. 2014. Durability of laterite/sand hollow blocks in magnesium sulphate environment. *American Journal of Engineering Research*, 03(08), 263–266.

Olutoge, F.A., Adeniran, K.M. and Oyegbile, O.B. 2014. The ultimate strength behaviour of laterised concrete beam. *Science Research 2013*, 1(3), 52–58.

Oner, A. and Akyuz, S. 2007. An experimental study on optimum usage of ggbs for the compressive strength of concrete. *Cement and Concrete Composites*, 29(6), 505–514

Parthiban, K., Saravanarajamohan, K., Shobana, S. and Bhaskar, A.A. 2013. Effect of replacement of slag on the mechanical properties of fly ash based geopolymer concretete. International Journal of *Engineering and Technology*, 5(3), 2555–2559.

Patil, Y.O. Patil, P.N. and Dwivedi, A.K. 2013. GGBS as partial replacement of OPC in cement concrete – an experimental study. *International Journal of Scientific Research*, 2(11), 189–191.

Rahman, M.A. 1988. Effect of rice husk ash on the properties of bricks made from fired lateritic soil-clay mix. *Materials and Structures*, 21(3), 222–227.

Ramalekshmi, M., Sheeja, R. and Gopinath, R. 2014. Experimental behaviour of reinforced concrete with partial replacement of cement with ground granulated blast furnace slag. *International Journal of Engineering Research & Technology*, 3(3), 525–534.

Rughooputh, R. and Rana, J. 2014. Partial replacement of cement by ground granulated blast furnace slag in concrete. *Journal of Emerging Trends in Engineering and Applied Sciences*, 5(5), 340–343.

Salau, M. A. and Balogun, L. A. 1998. Shrinkage deformations of laterized concrete. *Building and Environment*, 34(2), 165–173.

Schellmann, W. 1994. Geochemical differentiation in laterite and bauxite formation. *Catena*, 21(2–3), 131–143.

Siddique, R. and Kaur, D. 2012. Properties of concrete containing ground granulated blast furnace slag (GGBFS) at elevated temperatures. *Journal of Advanced Research*, 3(1): 45–51. DOI:10.1016/j.jare.2011.03.004.

Sumadi, S.R., Hamir, A.M.R. and Diah, A.B.M. 1999. Capability of GGBS concrete exposed to sea water. *Malaysian Science & Technology Congress*.

Supraja, V. and Kanta Rao, M. 2013. Experimental study on GeoPolymer concrete incorporating GGBS. *International Journal of Electronics, Communication & Soft Computing Science and Engineering*, 2(2), 71–78.

Sunil, B.M. and Krishnappa, H. 2012. Effect of drying on the index properties of lateritic soils. *Geotechnical and Geological Engineering*, 30, 869–879.

Udoeyo, F.F., Iron, U.H. and Odim, O. O. 2005. Strength performance of laterized concrete. *Construction and Building Materials*, 20(10), 1057–1062.

Ukpata, J.O. et al. 2012. Compressive strength of concrete using lateritic sand and quarry dust as fine aggregate. *ARPN Journal of Engineering and Applied Science*, 7(1), 81–93.

Udoeyo, F.F. Brooks, R. Udo-Inyang, P. and Iwuji, C. 2010. Residual compressive strength of laterized concrete subjected to elevated temperatures. *Research Journal of Applied Sciences, Engineering and Technology*, 2(3), 262–267.

Tardy, Yves. 1997. *Soils*. Spain, 105–110.

BIBLIOGRAPHY

Abhilash, D.A., William, B. and Rukesh, A.R. 2014. An experimental study on behavior of concrete by using combination of GGBS with rice husk ash and comparing it with the conventional concrete. *IOSRD International Journal of Engineering*, 1(2).

Ali, S.A. and Abdullah, S. 2014. Experimental study on partial replacement of cement by flyash and GGBS. *International Journal for Scientific Research & Development*, 2(07). ISSN: 2321-0613.

Anifowose, A.Y.B. 2000. Stabilisation of lateritic soils as a raw material for building blocks. *Bulletin of Engineering Geology and the Environment*, 58, 151–157.

Bowles, J.E. 1990. *Physical and Geotechnical Properties of Soil* (2nd edition). Mc Graw-Hill, Inc., 478.

Brightson, P.M., Premanand and Ravikumar, M.S. 2014. Flexural behaviour of beams incorporating GGBS as partial replacement of fine aggregate in concrete. *Advanced Materials Research*, 984, 698–706.

Gambhir, M.L. and Jamwal, Neha 2014. *Lab Manual – Building and Construction Materials (Testing and Quality Control)*. McGraw Hill Education (India) Private Limited, New Delhi.

Ganesha, A.V., Krishnaiah, C. and Jayappa, K.S. 2007. Mineralogical studies of laterites around Mangalore, Karnataka, India. *Indian Mineralogist*, 41(1), 76–83.

IS: 10050-1981. Method for determination of slake durability index of rocks. *Bureau of Indian Standards*, New Delhi.

IS: 10262-2009. Concrete mix proportioning – Guidelines. *Bureau of Indian Standards*, New Delhi.

IS: 1124-1974. Method of test for determination of water absorption of natural building stones. *Bureau of Indian Standards*, New Delhi.

IS: 2386-1963. Methods of tests for aggregates for concrete. *Bureau of Indian Standards*, New Delhi.

IS: 2720-1985a. Methods of test for soils: Part 4 grain size analysis. *Bureau of Indian Standards*, New Delhi.

IS: 2720-1985b. Methods of test for soils: Part 5 determination of liquid and plastic limit. *Bureau of Indian Standards*, New Delhi.

IS: 2720-1987. Methods of test for soils (Part 16) Laboratory determination of CBR. *Bureau of Indian Standards*, New Delhi.

IS: 383-1970. Specifications for coarse and fine aggregates from natural sources of concrete. *Bureau of Indian Standards*, New Delhi.

IS: 456-2000. Code of practice for plain and reinforced concrete. *Bureau of Indian Standards*, New Delhi.

IS: 516-1959. Methods of test for strength of concrete. *Bureau of Indian Standards*, New Delhi.

IS: 5816-1970. Splitting tensile strength of concrete – Method of test. *Bureau of Indian Standards*, New Delhi.

Madu, R.M. 1980. The performance of lateritic stones as concrete aggregates and road chippings. *Material of Construction*, 13(6) 403–411.

Mathew, G. and Paul, M.M. 2013. Influence of fly ash and GGBFS in laterized concrete exposed to elevated temperatures. *Journal of Materials in Civil Engineering*, 26(3), 411–419.

Mohamed, H.A. 2011. Effect of fly ash and silica fume on compressive strength of self-compacting concrete under different curing conditions. *Ain Shams Engineering Journal*, 2(2), 79–86.

Nogami, J.S. and Villibor, D.F. 1991. Use of lateritic fine-grained soils in road pavement base courses. *Geotechnical & Geological Engineering*, 9(3–4), 167–182.

Ola, S.A. 1980. Mechanical properties of concretionary laterites from rain forest and Savannah zones of Nigeria. *Bulletin of the International Association of Engineering Geology*, 21(1), 21–26.

Olufemi, O. 1986. Basic index properties, mineralogy and microstructure of an amphibolites derived laterite soil. *Bulletin of the International Association of Engineering Geology*, 33(1), 19–25.

Osinubi, K.J., Bafyau, V. and Eberemu, A.O. 2009. Bagasse ash stabilization of lateritic soil. *Appropriate Technologies for Environmental Protection in the Developing World*, 271–280.

Serafimovski, T., Volkov, A.V., Boev, B. and Tasev, G. 2013. The Rganovo Fe–Ni deposit: an example of the reworked lateritic weathering crust in the vardar ophiolite zone. *Doklady Earth Sciences*, 452(1), 899–903.

Shetty, M.S. 2003. *Concrete Technology - Theory and Practice*. S. Chand & Company Ltd., New Delhi.

Shetty, M.S. 2005. *Concrete Technology*. S. Chand & Company Ltd., New Delhi.

Siddharth and Munnur, S. 2015. Experimental study on strength properties of concrete using steel fibre and GGBS as partial replacement of cement. *International Journal of Engineering Research & Technology*., 4(01), 436–440.

5 Utilization of Coal Mine Waste in Vegetation

Vikas Chaitanya and Ram Chandar Karra

CONTENTS

DOI: 10.1201/9781003268499-5

LEARNING OBJECTIVES

- To understand the suitability of coal mine waste [overburden (OB) and ash] for vegetation
- To understand and determine the required characteristics of coal mine waste for vegetation
- To select the suitable species which can grow in mine waste material

5.1 INTRODUCTION

Open cast coal mining involves the extraction of coal deposits and the removal of overlying soil and rock, which is overburden dumps. The nature of the overburden dumps spoil may vary from relatively hard, coarse, un-weathered rock fragments to fine, soft, easily erodible earthy materials. In all cases, the common thread is the need to first remove whatever soil cover exists before mining activity can commence. This soil cover may either have been lost, or at least if stockpiled for later use in rehabilitation work is severely disturbed through mixing of topsoil with other waste rock. This results in the change of top soil quality and becomes challenging for the reclamation work of coal mine overburden dumps. Further, the impacts of overburden dumps are as follows: change in natural land topography, affects the drainage system and also prevents natural succession of plant growth resulting in acute problems of soil erosion and environmental pollution (Singh et al. 1994; Singh et al. 1996; Wali 1987). The end result for mining activities on the surface is the creation of mining waste, alteration of land forms which is a concern to the society and it is desired that the pristine conditions are restored. The measures commonly employed in the mined-out areas are: compensatory afforestation/plantation, reclamation, rehabilitation based on the existing mine closure guidelines. Successful and long-term mine spoil reclamation of overburden dumps requires the establishment of stable nutrient cycles from plant growth and microbial processes (Lone et al. 2008; Kavamura and Espostio 2010; Sheoran et al. 2010) and the selection of appropriate species for vegetation.

Mine reclamation and ecological restoration have become very important as a part of sustainable development strategy for the mining industry in many countries. Many case studies have been undertaken on mine reclamation addressing the soil fertility, structure, microbial populations, top soil management and nutrient cycling in order to retain the land to beneficial and as self-sustaining ecosystem.

Establishment of revegetation on mine overburden dumps renders biologically productive, contributing towards physical stabilization, pollution control, visual improvement and removal of threats to surrounding population. Overcoming the constraints to plant establishment can be facilitated through amelioration of physicochemical properties of dumped wastes and by choice of appropriate plant species.

The reclamation rate not only depends on one time vegetation/revegetation by using some species of trees and grasses/legumes and its survival rate but also needs the regular monitoring of the soil condition, species selection and dump characteristics. Regular monitoring helps us not only in the selection of plant species which is very important in ecorestoration process but also creation of locality-based micro-ecosystem.

5.1.1 Vegetation on Mine Dumps

Coal discard dumps are to be properly reclaimed and restored; otherwise, it may lead to many environmental problems which include erosion and sedimentation into surrounding water resources, contamination of surface and ground waters by acidic leachates and runoff damage from landslides due to failures of slopes and spontaneous combustion. These discarded spoils are not suitable for both plant and microbial growth because of low organic matter content, unfavourable pH and drought arising from coarse texture or oxygen deficiency due to compaction (Agarwal et al. 1993). The effects of mine wastes can be multiple, such as soil erosion, air and water pollution, toxicity, geo-environmental disasters, loss of biodiversity and ultimately loss of economic wealth (Wong 2003).

Productivity of mine waste including soil can be increased by adding various natural amendments, such as saw dust, wood residues, sewage sludge and animal manures, as these amendments stimulate the microbial activity which provides the nutrients (N, P) and organic carbon to the soil (Sheoran et al. 2010).

Benefits of compost to soils and vegetation are well researched. Compost provides organic matter; decreases bulk density and erosion; increases aggregate stability, aeration, water infiltration and retention (Tester 1990; Serra-Wittling et al. 1996; Ros et al. 2001). It increases concentrations and availability of micro and macro nutrients (Guerrero et al. 2001; Martínez et al. 2003), providing a wider range of nutrients than inorganic fertilizers with less nitrate leaching and water contamination (Gagnon et al. 1997; Mamo et al. 1999).

In fact, properties of mine waste can be altered either in-situ or ex-situ. In-situ alteration involves addition of additives or fertilizers that increases the physico-chemical properties of soil at the mine site. Whereas ex-situ alteration includes altering the properties of mine waste away from the site.

The use of coal ash and clean coal combustion by-products as soil additives or soil amendments to enhance establishment and growth of vegetation on coal mined lands as well as agricultural soils has been extensively studied (Adams et al. 1972; Capp 1978).

According to literature reviews by Carlson and Adriano (1993), the main benefits of using fly ash on mined lands are increased alkalinity and improved water-holding capacity (Stewart and Daniels 1995). While coal combustion by-products can alleviate plant growth limiting factors in mine spoils such as low pH and coarse texture, plant growth may still be limited by other factors. Coal combustion by-products are poor sources of nitrogen and phosphorus (Carlson and Adriano 1993).

Bio solids have proven effective in the reclamation and treatment of old mining sites. They are able to cost efficiently establish a vegetative cover on contaminated lands and limit the movement of metals through erosion, leaching and wind. A cap is formed upon the application of bio solids because of their permeability and water adsorption characteristics, which prevent water contact with contaminates in the soil below. Depending on the amendments added, bio solids can serve many purposes, including pH control, metal control and fertilization. Their adaptability allows them to conform to the specific characteristics of any reclamation site (Jenness 2001).

Logically ecorestoration of overburden should be done through natural succession process which commences with sowing of seeds of legumes, grasses, herbs and shrubs in the inter-spacing of tree plantation (IBM 2000; Maiti et al. 2002).

5.1.2 Mine Spoil Quality and Limiting Factors for Reclamation and Revegetation

Soil quality in simple terms can be defined as 'the capacity of soil to function' (Karlen et al. 1997), but it is difficult to measure and quantify. It is a broad concept that involves biological, chemical and physical properties that sustain productivity, environmental quality and support healthy organisms (Doran et al. 1996; Van Bruggen and Semenov 2000).

The long-term success of land reclamation efforts requires the maintenance and improvement of soil quality in terms of its physical, biological and chemical properties. These properties are proposed as basic indicators or part of a minimum data set of soil qualities that include both tilth, soil organic matter (SOM), total organic nitrogen and carbon, aggregate stability, aeration, macroscopy, water holding capacity, microbial biomass, mineralizable C and N, bulk density, resistance to erosion, nutrient retention capacity and pH and electrical conductivity (EC) (Doran and Parkin 1996; Larson and Pierce 1994; Karlen et al. 1992). Larson and Pierce (1994) observed SOM as one of the most important soil properties that contributes to soil quality and stability. Soil organic carbon (SOC) was also designed as a master variable to estimate soil attributes such as cation exchange capacity (CEC), water retention characteristics and leaching potential (Doran and Parkin 1996).

5.2 FACTORS INFLUENCING SOIL QUALITY

Reclamation and revegetation of abandoned coal lands are often limited by physical and chemical properties existing in the soil. Some of the factors that influence the soil quality and the vegetation process have been discussed in the following sections.

5.2.1 Soil Texture

Relative amount of sand (2.0–0.05 mm), silt (0.05–0.002 mm) and clay-sized (<0.002 mm) particles determine the texture of soil. The fraction of the soil 2 mm or larger has an impact on the behaviour of the whole soil. Soil properties, such as available water capacity, cation-exchange capacity, saturated hydraulic conductivity, structure and porosity are affected by the volume, composition and size distribution of fragments in the soil. Particle size influences pore space, which affects soil solution holding capacity and availability, microbiotic habitat quality, gas exchange and

TABLE 5.1
Soil Particle Size Classifications (USDA Soil Textural Classification System)

Name of Soil	Diameter Limits (mm)
Very coarse sand	2.0–1.0
Coarse sand	1.0–0.50
Medium sand	0.50–0.25
Fine sand	0.25–0.10
Very fine sand	0.10–0.05
Silt	0.05–0.002
Clay	Less than 0.002

buffering ability. For example, soils with very high amounts of sand (particle size 2.0 to 0.05 mm) typically give large pore spaces and high drainage rates but poor fertility. Soil clay particles and organic matter aid in aggregate stability, creating a macrostructure within the soil profile and enhancing chemical and biological activity (Lado et al. 2004).

Clay can increase the amount of water and nutrients stored. By slightly slowing the rate of water movement, it can reduce the rate of nutrient loss through leaching. If the amount of clay is great, it can impede water and air movement, restrict root penetration, increase runoff and, on sloping land, result in increased erosion (USDA Technical Handbook). Table 5.1 shows the USDA soil textural classification system.

5.2.2 Bulk Density and Soil Compaction

The most common measure of soil compaction is bulk density. Compaction of the growing media has the greatest impacts on the success of reforestation on reclaimed mined lands. Compaction increases bulk density and resistance to mechanical penetration (Barnhisel 1988).

Soil compaction occurs when moist or wet soil particles are pressed together and the pore spaces between them are reduced. Adequate pore space is essential for the movement of water, air and soil fauna through the soil. The mechanical strength and poor oxygen supply of compacted soil restrict root penetration. Soil moisture is unavailable if layers of compacted soil restrict root growth.

It has been recognized that excessive levels of these physical properties tend to reduce root growth, lowering the potential for successfully growing trees on reclaimed sites (Graves et al. 1995). The level of compaction of the growing medium is a function of physical properties, its moisture content, backfilling method, equipment type and number of passes of working.

Bulk density displays a strong correlation with tree survival rate (Conrad 2002). Compaction restricts infiltration, resulting in excessive runoff, erosion (Pierce et al. 1983), nutrient loss and potential water-quality problems. The forest area which consists of fine texture soils possesses less bulk density values and restricts runoff and erosion problems. According to Kostadinov and Marković (1996), the water-absorbing capacity of the forest cover is about 400% bigger than its mass in dry condition. Due to their high retention capacity, the forest ecosystems reduce runoff from 38 to 100 times.

Generally, the productive nature soils have bulk densities in the range of 1.1–1.5 gm/cm^3. Higher bulk density limits the rooting depth and also the perennial growth. Severely compacted (bulk density greater 1.7 gm/cm^3) mine soils, particularly those with less than 2 feet of effective rooting depth, shallow intact bedrock and the presence of large boulders in the soil simply cannot hold enough water to sustain plant growth (Sheoran et al. 2010).

5.2.3 Moisture Content and Water Holding Capacity

A critical environmental factor to plant growth is maintenance of adequate moisture in the root zone. The water is needed not only as a transport medium in which nutrients are absorbed and moved through the plant, and wastes are expelled, but more critically on a short-term basis, as a coolant. Transpiration of water by the plant is required to maintain non-lethal temperatures. Lack of water, for even a short time, has profound effects. Since plants absorb nearly all of their water through their roots, maintenance of adequate soil moisture in the root zone is essential.

Moisture content in a dump is a fluctuating parameter which is influenced by the time of sampling, height of dump, stone content, amount of organic carbon and the texture and thickness of little layers on the dump surface (Donahue et al. 1990). During winter, the average moisture content of 5% was found to be sufficient for the plant growth (Sheoran et al. 2010).

Soil water holding capacity and water availability are vital to successful revegetation on disturbed lands. Reclaimed soil texture and sufficient depth of rooting medium are two important factors in ensuring plants have adequate available water in a revegetation project (Bell 2002). Cover soil with sandy or coarse textures often has poor water holding capacity.

5.2.4 Soil pH and Electrical Conductivity

Soil pH is an indicator of mine soil quality. The mine soil pH depends upon the type of rock fragments present, SOM (plantation) and on other physical features. If the dump contains sulphide minerals, it results in low soil pH due to weathering and oxidation of sulphide minerals and creates toxicity problem in mine soils (Bradshaw 1997); on the other hand, high pH results with the carbonate ($Ca/MgCO_3$) bearing minerals and rocks as they tend to weather and dissolve.

Gitt and Dollhopf (1991) and Gould et al. (1996) suggest that a pH in the range of 6.0 to 7.5 is ideal for forages and other horticultural and agronomic uses. When the soil pH is below 5.54, reduced legume and forage growth occur due to metal toxicities such as aluminium and manganese, phosphorous fixation, and reduced population of N-fixing bacteria. This growth hence inhibits plant root growth and many other metabolic processes (Sheoran et al. 2010).

pH influences the soil structure formation, root growth and nutrient availability for the vegetation on the mine soil. Low soil pH or acidic soils have several adverse effects including Al and Mn toxicity and nutrient deficiencies (Bradshaw 1997; Sheoran et al. 2010).

Alkaline soils have a pH greater than 7 due to the presence of calcium carbonate from a calcareous parent geological material. Calcium carbonate buffers the pH from 7.5 to 8.5 (Marschner 1995). The main characteristic of a saline soil is the limited plant available water caused by ions dissolved in the soil solution (Hausenbuiller 1985).

EC is a common mine variable influencing plant productivity (Andrews et al. 2002). High levels of soluble salts inhibit water and carbon dioxide uptake and also inactivate enzymes affecting protein synthesis, C metabolism and photophosphorylation (Taiz and Zeiger 1991). Soil salinity reduces the availability of soil water to plants by increasing the soil-water potential, in particular the osmotic potential. This process stresses plants reducing or stunning growth. Both growth rate and size decrease as salinity increases (Jurinak and Suarez 1990).

High concentrations of sodium combined with low concentrations of salts cause soil aggregates to break down, reducing pore size, increasing bulk density and decreasing total porosity (Tisdale et al. 1993). Sodic soils have nutrient limitations and are deficient in iron, phosphorus and occasionally calcium, potassium and magnesium (Marschner 1995).

It's normal tendency to use alkaline or acidic specific plant species to increase productivity of soil. Gupta and Abrol (1990) reported that the productive capacity of alkaline soils has been found to improve by growing plants adapted to sodic soils. Reclamation agroforestry systems also have been noticed and the improvement in biological production and ameliorating sodic conditions of soils has been occurred with the increasing SOM content and availability of soil inorganic nitrogen (Singh et al. 1994; Singh et al. 1996).

5.2.5 Trace Metals and Micronutrients Availability

Trace elements, specifically metals, are found in ore bodies and released into the environment during mining, milling and smelting processes. These elements often create toxicity problems in soils and contaminate surface and ground waters, creating exposure risks to humans, wild life and aquatic organisms. Once soils are contaminated with metals, metal levels are relatively static and cannot be removed by natural processes (Bradshaw 1997). Chemical properties of mine wastes are considered the greatest restraint to plant growth. Trace elements considered essential for plant growth include B, Ca, Co, Co, Fe, Mn, Mo, Si, Se and Zn. The important metallic micronutrients that are essential for plant growth are Fe, Mn, Cu and Zn.

5.2.6 Soil Organic Carbon and Nutrient Content (N, P and K)

SOM increases water holding capacity, soil porosity, infiltration and CEC (Sheoran et al. 2010; Dickinson 2002).

Mining and reclamation activities cause drastic loss of the antecedent soil organic carbon (SOC,N70%) and nitrogen (N,N65%) in mined and newly reclaimed mine soils (Akala and Lal 2001).

Three major macronutrients, namely nitrogen, phosphorus and potassium, are generally found to be deficient in overburden dumps (Coppin and Bradshaw 1982; Sheoran et al. 2010). All these nutrients generally enter the soil solution from decomposition of organic matter soil minerals, atmospheric deposition and symbiotic mycorrhizal associations.

A level of organic carbon greater than 0.75% indicates good fertility (Ghosh et al. 1983). Maiti and Ghose (2005) reported the level of organic carbon to be 0.35% to 0.85% in the mine overburden. Nitrogen levels up to 1000 Kg/ha may be needed on reclaimed land to overcome the amount that would be provided by decomposing organic matter (Dickinson 2002). The deficiencies of nitrogen can be overcome by the establishment of nitrogen fixing plants and biological fixation.

Plant available phosphorous level increases with high levels of decomposing organic matter due to the release of phosphorous during the decomposition of organic matter. Phosphate availability is limited in acidic and alkaline soils (Dickinson 2002). The presence of phosphate may reduce the toxicity of lead, zinc and copper through precipitation and ion competition reactions (Johnson and Williamson 1994).

A concentration level of 11–25 kg/ha indicates medium fertility while concentrations of greater than 25 kg/ha indicate higher fertility for the mine soil for reclamation (Maiti and Ghose 2005).

Soil available potassium (K) is majorly derived from the mineral fraction of soil. Potassium uptake from plant roots is related to the concentration gradient between soil and root, rate of K diffusion through soil to root surfaces and root surface area. Soil moisture is a driving force for potassium uptake, and as soil dries, uptake becomes increasingly difficult (Foth and Ellis 1997).

A rating level of 120 to 280 kg/ha and greater than 280 kg/ha indicates medium fertility and higher fertility for the mine soil quality to be reclaimed (Maiti and Ghose 2005).

SOC is one of the most important constituents of the soil due to its capacity to affect plant growth as both a source of energy and a trigger for nutrient availability through mineralization. An increase in SOM leads to greater biological diversity in the soil, thus increasing biological control of plant diseases and pests.

5.2.7 Cation Exchange Capacity and Exchangeable Cations (Na, K, Ca, Mg)

Cation-exchange capacity is a measure of the ability of a soil to retain cations, some of which are plant nutrients. Soils that have a low cation-exchange capacity hold fewer cations and may require more frequent applications of fertilizer than soils that have a high cation-exchange capacity. Soils that have a high cation-exchange capacity buffer fluctuations in nutrient availability and soil pH.

Soil is made up of many components. A significant percentage of many soils is clay, organic matter, while a small percentage of many minerals is also important for several reasons. Both of these soil fractions have a large number of negative charges on their surface, thus they attract cation elements and contribute to a higher CEC. At the same time, they also repel anion ('like' charges).

Potassium regulates the opening and closing of the stomata by a potassium ion pump. Since stomata are important in water regulation, potassium reduces water loss from the leaves and increases drought tolerance. Exchangeable potassium in overburden soil sample is 53.336 meq/100 gm. The predominant role of magnesium is as a major constituent of the chlorophyll molecule, and it is therefore actively involved in photosynthesis. It is a co-factor in several enzymatic reactions that activate the phosphorylation processes and also assists the movement of sugars within a plant.

Magnesium is the key element of chlorophyll production and acts as an activator and component of many plants enzymes.

5.2.8 Trace Metals and Micronutrients Availability

Trace elements, specifically metals, are found in ore bodies and released into the environment during the mining, milling and smelting processes. These elements often create toxicity problems in soils and contaminate surface and ground waters, creating exposure risks to humans, wild life and aquatic organisms. Once soils are contaminated with metals, metal levels are relatively static and will not be removed by natural processes (Bradshaw 1997). Chemical properties of mine wastes are considered the greatest restraint to plant growth. Trace elements considered essential for plant growth include B, Ca, Co, Cu, Fe, Mn, Mo, Si, Se and Zn. The important metallic micronutrients that are essential for plant growth are Fe, Mn, Cu and Zn.

Trace elements considered essential for plant growth include B, Ca, Co, Cu, Fe, Mn, Mo, Si, Se and Zn. The important metallic micronutrients that are essential for plant growth are Fe, Mn, Cu and Zn. There are about seven nutrients essential to plant growth and health that are only needed in very small quantities. These are manganese, boron, copper, iron, chromium, cadmium and zinc. Though these are present in only small quantities, they are all necessary.

Copper is a component of some enzymes. Symptoms of copper deficiency include browning of leaf tips and chlorosis. Manganese activates some important enzymes involved in chlorophyll formation. Manganese deficient plants will develop chlorosis between the veins of its leaves. The availability of manganese is partially dependent on soil pH. Zinc participates in chlorophyll formation and also activates many enzymes. Symptoms of zinc deficiency include chlorosis and stunted growth.

5.2.9 Energy-Dispersive X-Ray Spectroscopy

Metal concentrations in soil range from less than 1 mg/kg to high as 100,000 mg/kg, whether due to the geological origin of the soil or as a result of human activity (Blaylock and Huang 2000). Excess concentrations of some heavy metals in soils such as Cd^{+2} Cr^{+6}, Cu^{+2}, Ni^{+2} and Zn^{+2} have caused the disruption of natural aquatic and terrestrial ecosystems (Gardea-Torresdey et al. 1996; Meagher 2000). Cadmium and lead, which have no known beneficial effects, may become toxic to plants and animals if their concentrations exceed certain values (Adriano 1986).

Energy-dispersive X-ray spectroscopy (EDS, EDX, or XEDS) is an analytical technique used for the elemental analysis or chemical characterization of a sample. It relies on the investigation of an interaction of some source of X-ray excitation and a sample. Its characterization capabilities are due in large part to the fundamental principle that each element has a unique atomic structure allowing unique set of peaks on its X-ray spectrum. To stimulate the emission of characteristic X-rays from a specimen, a high-energy beam of charged particles such as electrons or protons, or a beam of X-rays, is focused into the sample being studied. Higher-energy shell and the lower energy shell may be released in the form of an X-ray. The number and energy of the X-rays emitted from a specimen can be measured by an energy-dispersive spectrometer. As the energy of the X-rays is characteristic of the difference in energy between the two shells, and of the atomic structure of the element from which they were emitted, this allows the elemental composition of the specimen to be measured.

Electron beam excitation is used in electron microscopes, scanning electron microscopes (SEM) and scanning transmission electron microscopes (STEM). X-ray beam excitation is used in X-ray fluorescence (XRF) spectrometers. A detector is used to convert X-ray energy into voltage signals; this information is sent to a pulse processor, which measures the signals and passes them onto an analyser for data display and analysis. The most common detector now is Si (Li) detector cooled to cryogenic temperatures with liquid nitrogen; however, newer systems are often equipped with silicon drift detectors (SDD) with Peltier cooling systems. Chemical characterization is done by using SEM by performing energy dispersive X-ray analysis (EDX).

5.3 EXPERIMENTAL INVESTIGATIONS

Coal mine waste samples were collected from an overburden dump of a coal mine in South India. Physico-chemical characteristics were analysed and additives were added to enhance its fertile characteristics.

5.3.1 Collection of Samples

Samples were collected from different locations of the dumps using random sampling technique. From each location, V-shaped pits of 50 cm x 50 cm dimension each were dug with 50 cm depth at each plot using spade and trowel, waste material was scraped from each of the 8 pits randomly chosen.

The difference in the values for pH and EC is less than 10% and pH varied between 7.67 and 8.1. The samples were mixed after determining pH, EC and also after getting analysed with SEM for heavy metal contamination.

Analysis was done to find the physico-chemical characteristics of the OB sample and found to have very poor nutrient content. After sieve analysis only the finer fraction (passing through 4.75 mm) was taken and additives were added in order to increase the fertile characteristics of OB sample.

The additives that were chosen were:

- Bottom Ash- From project site area which is being used for stowing.
- Secondary sludge from sewage treatment plant.
- Fly ash from thermal power plant.
- Limestone powder.

These additives were added in the ratio of 25% V/V and 50% V/V with respect to OB sample.

5.3.2 Texture Analysis

Texture (or particle size) analysis is a procedure to determine the relative proportions of the different particle (or grain) sizes which make up a given soil mass. As explained by Punmia et al. (2005) two techniques are used in this exercise to separate the soil particles into particle-size ranges, namely clay (<0.002 mm), silt (0.002–0.05 mm) and sand (0.05–2 mm). Coarse particles, sands and gravels have been analysed with mechanical sieves. The distribution of fine particle sizes (silts and clays) are determined by uniformly dispersing the soil in water measuring how quickly the particles fall in the mixture, the sedimentation analysis.

Table 5.2 was obtained from results of particle-size analysis; it reveals that the major part of the samples collected is loamy sand to sandy loam in nature.

Table 5.2 shows the proportions of sand, silt and clay percentages in OB sample and in additives. Fly ash mixed with OB sample in 50% V/V has the highest percentage of clay (4.4%) and the highest percentage of silt (36.18%). The OB sample has the least percentage of clay (0.42%) and silt (7.44 %) than any other additives and is sandy in nature.

5.3.3 Sieve Analysis

Sieving is performed by arranging various sieves one over the other in the order of their mesh opening size, the largest aperture sieve being kept on the top and the smallest at the bottom. Sieve analysis results are shown in Table 5.3.

TABLE 5.2
Proportions of Sand, Silt and Clay (Finer Fraction) in the Overburden (OB) Sample and Additives

S. No.	Sample	(<2 μm) Clay %	(2–50 μm) Silt %	(50 μm–2 mm) Sand %	Texture Class
1	OB sample	0.42	7.44	86.54	Sand
2	Bottom Ash 25% V/V	1.27	14.44	81.65	Loamy sand
3	Bottom Ash 50% V/V	1.49	19.58	75.56	Loamy sand
4	Secondary sludge from STP 25% V/V	2.23	18.34	71.65	Loamy sand
5	Secondary sludge from STP 55% V/V	2.48	33.48	61.79	Sandy loam
6	Fly Ash 25% V/V	1.60	24.42	67.36	Sandy loam
7	Fly Ash 50% V/V	2.21	36.18	52.11	Sandy loam
8	Limestone 25% V/V	3.20	22.76	70.28	Loamy sand
9	Limestone 50% V/V	3.80	29.65	63.96	Sandy loam

TABLE 5.3
Sieve Analysis Values for OB Sample

IS Sieve Size (mm)	Weight Retained (gm)	% of Weight Retained	Cumulative % Retained	Cumulative % Passed
4.75	0	0	0	100
2.36	56.7	5.67	5.67	99.44
1.18	153.5	15.35	21.02	95.72
0.6	391.5	39.15	60.17	91.9
0.3	317.3	31.73	91.9	60.17
0.15	38.2	3.82	95.72	21.02
0.075	37.2	3.72	99.44	5.67
Pan	5.6	0.56	100	0
	Total=1000	Total=100		

5.3.4 Field Bulk Density and Moisture Content Analysis

The water or moisture content of the field sample was determined in the quality control laboratory of the mining company by oven dry method according to BIS: 2720(part II)-197. The field bulk density was also determined by core cutter method according to HIS: 2720 (part XXIX)-1973/88.

Generally, the productive natural soils have bulk densities in the range of 1.1 gm/cc to 1.5 gm/cc (Kumar 2007). For many plants, a soil bulk density more than 1.5 gm/cc in fine-textured soils can restrict root growth (Kozlowski 1999). Table 5.4 shows general relationship of soil bulk density to root growth based on soil texture (Pierce et al. 1983).

TABLE 5.4
General Relationship of Soil Bulk Density (gm/cc or kg/m³) to Root Growth based on Soil Texture (Pierce et al. 1983)

Soil Texture	Ideal Bulk Densities (gm/cc)	bulk Densities that May Affect Root Growth (gm/cc)	Bulk Densities that Resists Root Growth (gm/cc)
Sands, loamy sands	<1.60	1.69	>1.80
Sandy loams, loams	<1.40	1.63	>1.80
Sandy clay loams, loams, clay loams	<1.40	1.60	>1.75
Silts, silt loams	<1.30	1.60	>1.75
Silt loams, silty clay loams	<1.40	1.55	>1.65
Sandy clays, silty clays, some clay loams (35–45% clay)	<1.10	1.49	>1.58
Clays (> 45% clay)	<1.10	1.39	>1.47

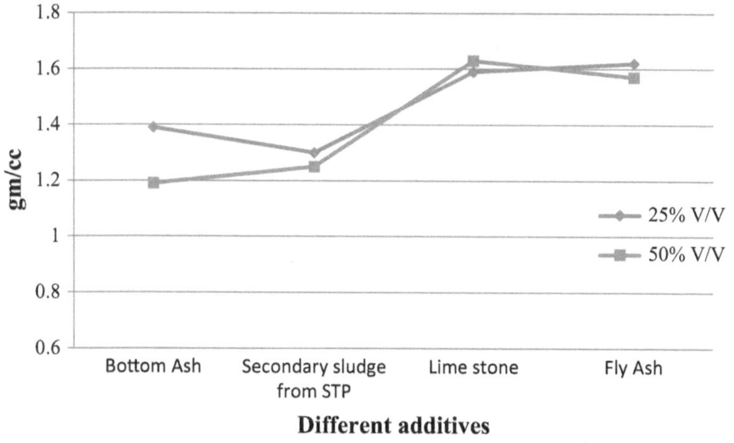

FIGURE 5.1 Variation of bulk density with different additives.

Figure 5.1 shows the variation of bulk density with different additives. Bottom ash mixed with OB sample in 50% V/V has the lowest value of bulk density. The bulk density goes on increasing with sewage sludge, fly ash and limestone powder, respectively. On addition of additives, bulk density was decreased from 1.76 gm/cc and ranged in between 1.19 gm/cc and 1.63 gm/cc. On addition of bottom ash to sample, the bulk density decreased to 1.19 gm/cc which was the least among other additives. Bottom ash mixed with OB sample in 50% V/V, and both the proportions of sewage sludge mixed with OB have bulk density values in desirable range for plant growth.

5.3.5 Determination of Moisture Content Using Oven Drying Method

The field samples collected in polythene bags were immediately brought down to the lab for the determination of moisture content by oven dry method. The average field moisture content was found to be around 4.14% of all the samples. Field moisture content in a dump is a fluctuating parameter which is influenced by the time of sampling, height of dump and amount of organic carbon, texture and the thickness of the litter layers of the dump surface (Donahue et al. 1990). The moisture content was bit more for sample 1, 2, 3 and 4 as they were collected during the end of monsoons in the month of September. So the moisture content is less in the samples collected later in November.

5.3.6 Soil pH and Electrical Conductivity

The pH is the measure of the intensity of the alkalinity or/and acidity of the soil suspension. It highly depends on soil water ratios and its determination according to BIS: 2720 part XIXX (1973). The EC of the soil sample is the measure of the soluble salts present. EC of the soil samples was measured using conductivity meter according to BIS: 2720 part XIX (1977).

pH of the OB sample was 7.72 and electrical conductivity of the OB sample is 0.36 ms. Changes in soil pH and EC are inevitable due to disturbance in soils, usually resulting in increased pH and EC. Figure 5.2 and Figure 5.3 show the pH and EC of different samples, respectively.

Initially the pH of OB sample was 7.72 but on addition of additives it varied from 6.27 to 9.14. In Figure 5.2, a large variation was seen on addition of fly ash in 50%v/v combination, whereas pH was decreased on addition of STP sludge in 25% V/V and 50% V/V combination. There can be presence of carbonaceous ($CaCO3/MgCO3$) sandstone and shale mineral fractions which are buffering the soil to pH 8. Munshower (1994) reported that high pH causes the calcium mobility and availability to be high, and hence the alkaline pH is observed in some soil samples. pH influences the soil

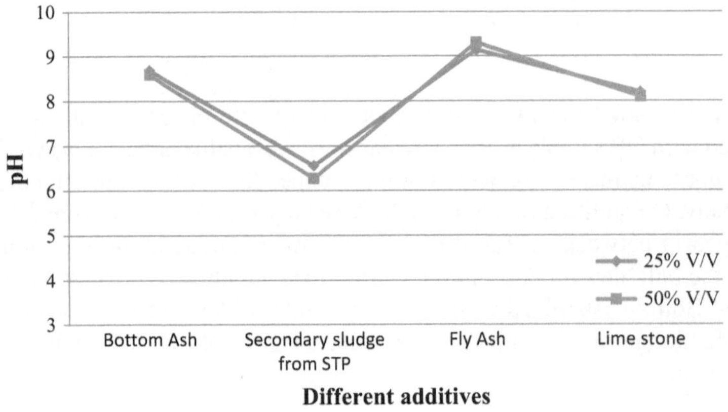

FIGURE 5.2 Variation of pH with different additives.

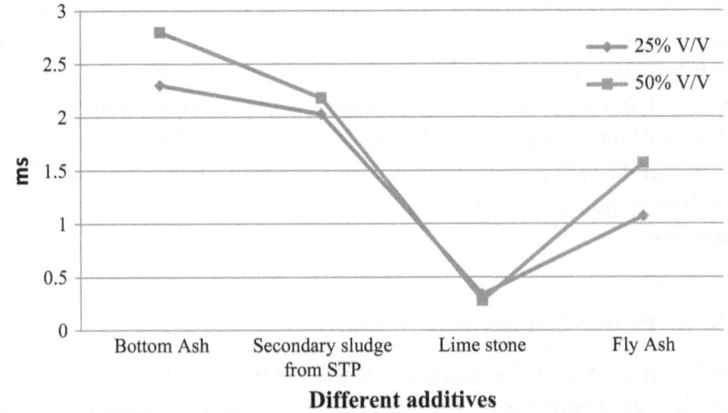

FIGURE 5.3 Variation of Electrical Conductivity (EC) with different additives.

structure formation, root growth and nutrient availability for the vegetation on the mine soil. Decrease in pH was observed for fly ash with time. After one month, the pH was decreased to 8.3. The pH of the additives of sewage sludge is in the favourable conditions for plant growth.

Initial EC of the OB sample is 0.36 ms. Figure 5.3 shows the variation of EC with different additives. Munshower (1994) said that soil EC is an integrated indicator for soil physical and chemical properties that are strongly related to crop yield. On addition of additives, the EC of mixture varies from 0.28 ms to 2.8 ms. Bottom ash mixed with OB sample showed maximum conductivity in 50%V/V combination. The EC values decreased almost with different additives from bottom ash being the highest and limestone being the least.

5.3.7 Water Holding Capacity

The water holding capacity was measured by gravimetric method (Trivedi et al. 1987). It is the maximum amount of water held in saturated soils. Soil water holding capacity and water availability are vital to successful revegetation on disturbed lands. Bell (2002) stated that reclaimed soil texture and sufficient depth of rooting medium are two important factors in ensuring plants have adequate available water in revegetation project.

Initial water holding capacity of OB sample is 17.52%. Figure 5.4 shows the variation of Water Holding Capacity with different additives. The plot between 25% and 50% V/V are almost parallel with each additive. The change in water holding capacity remained in small range on addition of additives. Water holding capacity was increased to 34% on addition of bottom ash to OB sample in 50% V/V mix from 17.52%. The least value of WHC was limestone mixed with OB in 25% V/V. Percentage of water holding capacity should be more for the better growth of plants.

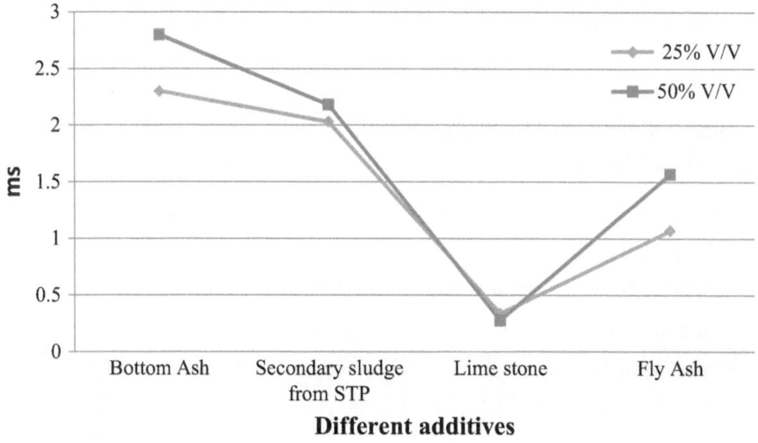

FIGURE 5.4 Variation of Water Holding Capacity with different additives.

5.3.8 Soil Organic Carbon Using Modified Walkley-Black Method

To determine the SOC, modified Walkley-Black Method is followed. This method involves the following steps.

i. A suitable quantity of soil sample (g) is weighed and transferred into a 500 mL conical flask.
ii. 10 mL 1N potassium dichromate and 20 mL of concentrated sulphuric acid are added slowly and shaken vigorously after adding, the mixture is kept idle for 30 minutes.
iii. 200 mL of distilled water and 10 mL phosphoric acid are added to the conical flask.
iv. FAS (0.4N) is taken in a burette and filled up to mark.
v. Three drops of diphenylamine indicator are added to the comical flask; the solution changes to blue and is titrated immediately against FAS till the end point is green in colour. Noted down the reading as mL of FAS consumed against sample solution (T).
vi. Prepared blank solution using the same procedure as above without soil sample in the conical flask and titrated against 0.4 N FAS till the end point is green in colour. Noted down the reading as mL of FAS consumed against blank solution(S).

Calculation: Percentage of organic carbon is calculated using the following expression:

$$\%C = \frac{3.951}{g}\left(1 - \frac{T}{S}\right)$$

T = Volume of FAS consumed against sample extract solution,
S = Volume of Fas consumed against blank solution,
g = Weight of the soil sample taken, gm.

Percentage of organic matter is calculated using the expression as given below:

$$\%OM = \%C \times 1.724$$

The values obtained by following the above-mentioned procedure are given in Table 5.5. Initial SOC for OB sample was found to be 0.18%

TABLE 5.5
Values of Soil Organic Carbon (%) for Different Additives

Additives/Mix %	25% V/V	50% V/V
Bottom ash	0.49	0.3
Secondary sludge from STP	2.75	3.58
Limestone	0.43	0.3
Fly ash	0.32	0.35

5.3.9 Total Nitrogen Using Kejhdhal Method

Total Nitrogen for the OB sample and additives was determined by Kejhdhal Method which includes digestion and distillation. The method involves the following steps.

Digestion:

i. Approximately 5 gm of soil sample is passed through a 2 mm sieve.
ii. About 1 gm of this material (accuracy 0.01 gm) is weighed into a digestion flask of soils rich in organic matter (>10%), else, 0.5 gm is weighed.
iii. 5–6 gm of k_2SO_4, a pinch of $CuSO_4$ and 10 mL of sulphuric acid-selenium digesting mixture are added. The solution is digested till we get white fumes and make up to 200 mL of sample using distilled water.

Distillation:

i. A known volume (200 mL) of the digest sample is taken in kjeldal apparatus and at the other end 0.1 N sulphuric acid is placed and 2–3 drops mixed indicator is added to the beaker.
ii. 5 mL NaOH 50% is added and distillation is done till 100 mL distillate is collected.
iii. After removing from flame, the lid is immediately opened. Then 100 mL sample is collected into conical flask and titrated against 0. N NaOH.
iv. For blank 25 mL of 0.1 N sulphuric acid, 2–3 drops of mixed indicator are added and titrated against 0.1 N NaOH.

Calculation: percentage of organic nitrogen is calculated using the following formula:

$$\%\text{TN} = \frac{a-b}{s} \times N_{NaOH} \times 1.4$$

where,

A = mL HCl required for titration blank,
B = mL HCl required for titration sample,
S = Air-dry sample weight in gram,
N_{NaOH} = Normality of NaOH (here, 0.1075), and 1.4 = 14 $\times 10^{-4} \times 100$ % (14 = atomic weight of nitrogen).

Nitrogen is a key element in plant growth. It is found in all plant cells, in plant proteins and hormones and in chlorophyll. Initial total Nitrogen of OB sample is 0.52%. Figure 5.5 shows that sewage sludge mixed in 50% V/V proportions with OB sample have almost 0.9% of total nitrogen. OB sample had higher percentage of total nitrogen than most of the additives. Only sewage sludge mixed with OB sample has higher percentage of total nitrogen than OB sample. Nitrogen plays an important role in the growth of plants. Higher nitrogen content is favourable for the growth of plants. Sewage sludge is the only additive having more total nitrogen than OB sample.

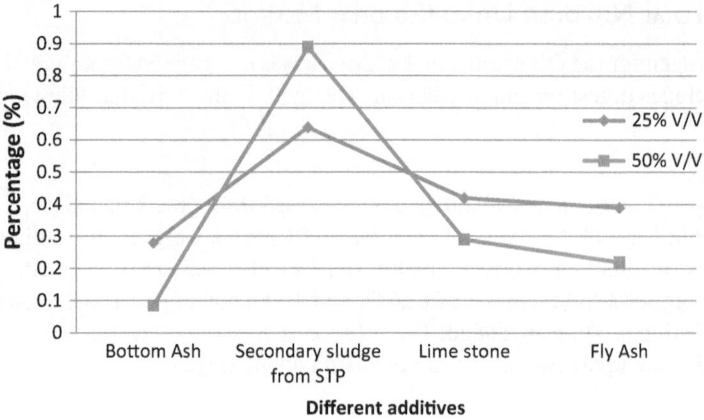

FIGURE 5.5 Variation of total Kjeldal nitrogen with different additives.

5.3.10 Available Phosphorous

Phosphorous in soils is generally determined as available phosphorous, which can be extracted by 0.002N sulphuric acid from the soil sample (Trivedi et al. 1987). This method involves the following steps:

Procedure:

i. Around 2 mm (5 gm) fresh air dried soil is taken in a 500 mL conical flask and 200 mL of 0.002N sulphuric acid is added.
ii. The suspension is shaken on rotary shaker for around half an hour.
iii. After shaking, the suspension is filtered through Whatman No.2 to get clear solution.
iv. 50 mL of the clear colourless solution is taken in a conical flask. If the sample is having colour and colloidal impurities, these are removed by adding a spoonful of activated charcoal and filtering.
v. 2 mL of ammonium molybdate solution and 5 drops of stannous chloride reagent are added. A blue colour appears in the presence of phosphate.
vi. Optical density reading is taken at 690 nm on a spectrometer using distilled water blank with same amount of chemical.
vii. A standard curve is prepared using the standard phosphate solution in the range 0 to 10.0 mg/L at the interval of 1 mg/L using spectrophotometer at 690 nm.

Calculation: The concentration of p in mg/L is found using the standard curve, while the percentage of available phosphorous is calculated from the following expression:

$$\%\text{available p} = \frac{mg\frac{P}{L}\ in\ the\ soil\ solution}{50}$$

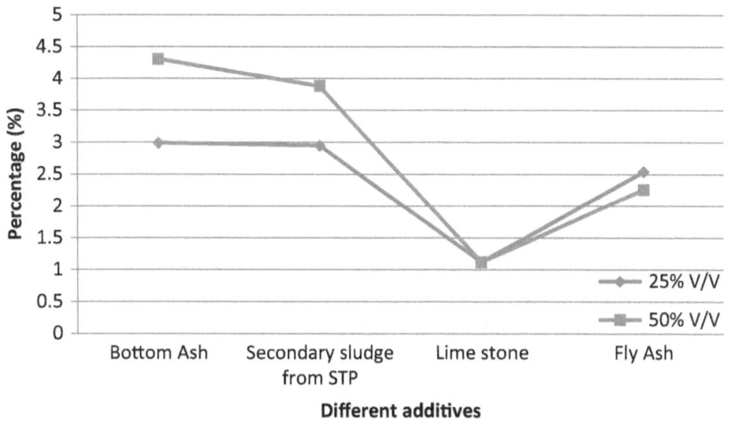

FIGURE 5.6 Variation of available phosphorus with different additives.

Phosphorus helps transfer of energy from sunlight to plants, stimulates early root and plant growth, and hastens maturity. Figure 5.6 shows the variation of available phosphorus with different additives. Available phosphorus of the OB sample is 2.67%.

From Figure 5.6, it is clear that the percentage of available phosphorus decreased from bottom ash, sewage sludge and limestone. There was no difference in the values of limestone mixed with OB sample in both 25% V/V and 50% V/V proportions. Limestone is having the least amount of phosphorus (1.12%), which is less than OB sample. Bottom ash mixed in 50%V/V proportion with OB sample has the highest percentage of available phosphorus. Bottom ash has the highest amount of available phosphorus which makes it favourable for plant growth with respect to available phosphorus.

5.3.11 Cation Exchange Capacity

Cation exchange is the physico-chemical process whereby one type of ions (cation) absorbed on soil particles is replaced by another type. The CEC signifies the capacity of soil to retain cations up to its highest limit; or it can also be defined as the soil to combine with cation in such a manner that they cannot be easily removed by leaching with water, but can be exchanged by an equivalent amount of other cations. It is the sum of exchangeable metallic ions plus hydrogen ions in a soil, and it is determined as per IS: 2720 (PATRT XXIV-1976).

Soils that have a low cation-exchange capacity hold fewer cations and may require more frequent applications of fertilizer than soils that have a high cation-exchange capacity. Initial CEC of OB sample is 31.99 meq. Figure 5.7 shows the Variation of CEC with different additives. CEC for OB sample after adding additives varied from 24.66 meq to 93.29 meq. The highest CEC was shown by OB sample mixed with limestone in 50% V/V which was almost twice the values of other additives. On addition of sewage sludge in 25% V/V, there was a decrease in CEC from 31.66 meq to 24.66 meq which may be due to very low amount of CEC of sewage sludge.

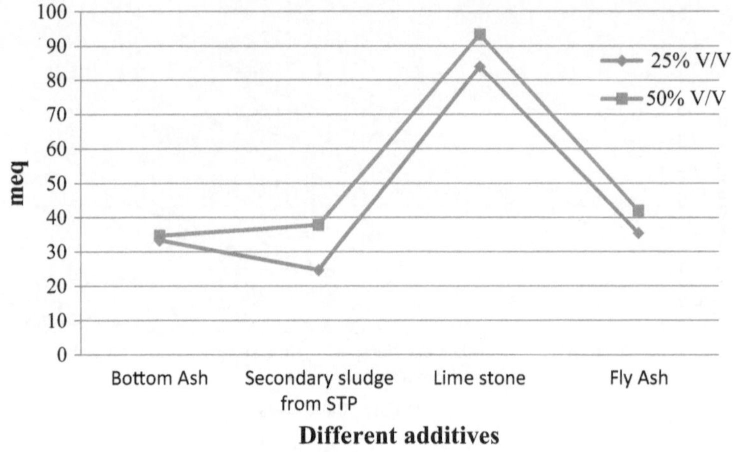

FIGURE 5.7 Variation of Cation Exchange Capacity with different additives.

5.3.12 Exchangeable Sodium and Potassium

The exchangeable sodium and potassium can be found by measuring the difference between extractable sodium and potassium values obtained from soil extract. The extract of sodium and potassium was determined by ammonium acetate method (Richards 1954). The extractable potassium is the total available potassium for a given soil sample. Exchangeable sodium and potassium in overburden soil sample is 40 meq/100 g and 53.336 meq/100 g, respectively.

In Figure 5.8, the highest quantity of exchangeable sodium is observed in sewage sludge (193.33 meq/100 gm) mixed with OB sample in 50% V/V proportion, which is almost five times of the initial value of OB sample. This shows that the sewage sludge is rich in sodium content.

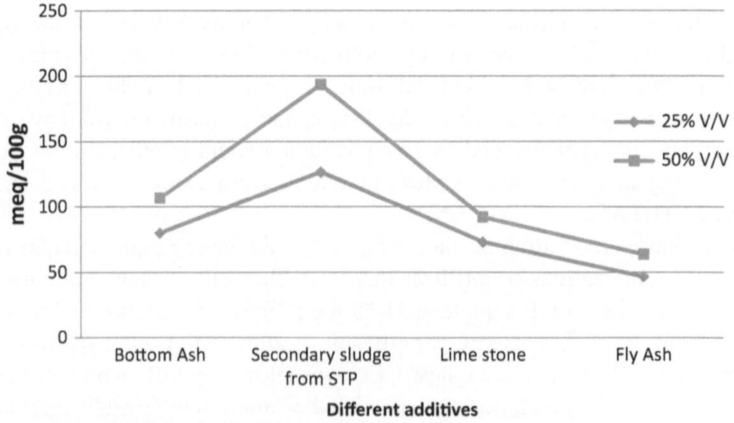

FIGURE 5.8 Variation of exchangeable sodium with different additives.

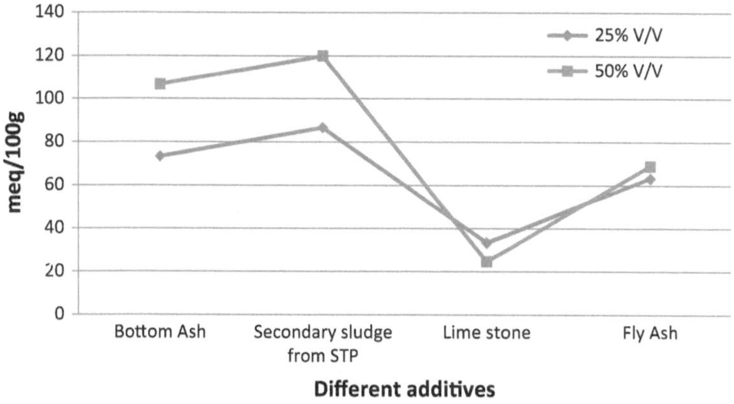

FIGURE 5.9 Variation of exchangeable potassium with different additives.

Potassium regulates the opening and closing of the stomata by a potassium ion pump. Since stomata are important in water regulation, potassium reduces water loss from the leaves and increases drought tolerance. Exchangeable potassium in overburden soil sample is 53.336 meq/100 g.

Figure 5.9 shows the exchangeable potassium ranging from 33.33 meq/100 gm to 120 meq/100 gm. Sewage sludge in 50% V/V with OB sample has the highest amount of potassium whereas limestone 50% V/V has the least. Even after addition of limestone in 25% V/V to OB sample showed lesser value of exchangeable potassium than OB sample. Another major nutrient required for growth of plants after nitrogen and phosphorus is potassium. Sewage sludge has higher amounts of potassium and is best suited for the growth of plants than other additives with respect to exchangeable potassium.

5.3.13 Exchangeable Calcium and Magnesium

Exchangeable calcium and magnesium of the OB sample and additives were determined as explained by Trivedi et al. (1987). Exchangeable calcium in overburden soil sample is 19.02 meq/100 g. Figure 5.10 shows that sewage sludge has the highest amount of exchangeable calcium. Overburden mixed with limestone 50% V/V has 88.96 meq/100 gm of exchangeable calcium whereas fly ash has very less amount of calcium.

The predominant role of magnesium is as a major constituent of the chlorophyll molecule, and it is therefore actively involved in photosynthesis. It is a co-factor in several enzymatic reactions that activate the phosphorylation processes and also assists the movement of sugars within a plant. Magnesium is the key element of chlorophyll production. It acts as an activator and component of many plant enzymes. Exchangeable magnesium in overburden soil sample is 24.8 meq/100 g.

From the Figure 5.11, exchangeable magnesium is the highest in OB sample mixed with sewage sludge in 50% V/V proportion and least in fly ash 50% V/V mix. Samples mixed with fly ash showed the lesser values of exchangeable magnesium

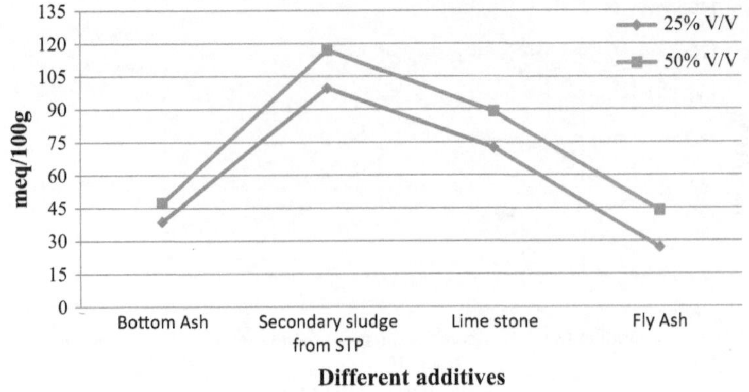

FIGURE 5.10 Variation of exchangeable calcium with different additives.

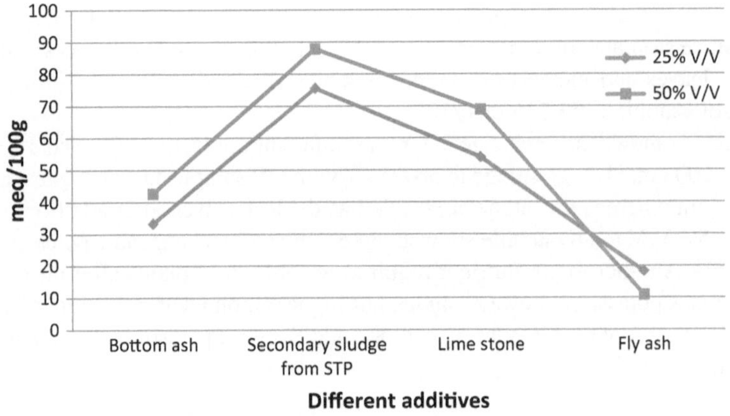

FIGURE 5.11 Variation of exchangeable magnesium with different additives.

than initial value of OB sample. This may be due to presence of exchangeable magnesium in very less percentage in fly ash.

5.3.14 Micronutrients (Fe, Zn, Mn, Cu, Cr and Cd) by DTPA Extraction Method

Micronutrients also play a major role in the growth of plants. They are found by DPTA extraction method as explained by Lindsay and Norvell (1978). Trace elements considered essential for plant growth include B, Ca, Co, Cu, Fe, Mn, Mo, Si, Se and Zn. The important metallic micronutrients that are essential for plant growth are Fe, Mn, Cu and Zn. There are about seven nutrients essential to plant growth and health that are only needed in very small quantities. These are manganese, boron, copper, iron, chromium, cadmium and zinc. Though these are present in only small quantities, they are all necessary. Table 5.6 shows the micronutrients present in additives

TABLE 5.6
Micronutrients Present in Additives and OB Sample

	Element (ppm)						
Sample	**B**	**Cu**	**Mn**	**Cd**	**Fe**	**Cr**	**Zn**
Bottom Ash 25% V/V	68.67	10.93	122.67	3.73	5424.03	7.533	30.66
Bottom Ash 50% V/V	54.67	14.13	122.67	1.33	6173.36	BDL	26.00
Secondary sludge from STP 25% V/V	94.33	35.93	77.33	0.47	5750.69	86.00	74.67
Secondary sludge from STP 55% V/V	116.67	65.60	99.33	0.33	7040.03	102.67	156.67
Limestone 25% V/V	8.33	9.80	80.00	BDL	7672.04	264.00	28.67
Limestone 50% V/V	6.33	12.13	99.33	BDL	9039.39	233.33	46.67
Fly Ash 25% V/V	18.67	10.27	80.00	0.18	2789.35	296.00	26.00
Fly Ash 50% V/V	19.33	10.80	97.33	4.33	2775.34	350.67	27.33
OB sample	12.67	5.80	70.00	BDL	4043.35	148.53	19.33

and OB sample. Iron is found in all the samples and in much higher concentrations than any other micronutrient which is essential for the synthesis of chlorophyll. The highest amount of iron can be found in limestone-OB 50% mixes and the least being fly ash-OB mixes.

Next important micronutrient is Boron (B), which is believed to be involved in carbohydrate transport in plants; it also assists in metabolic regulation. It is essential for seed and cell wall formation. Sewage sludge-OB mix in 50% V/V combination has the highest amount of boron where fly ash-OB 50% mixes have least amount. Copper is a component of some enzymes. Symptoms of copper deficiency include browning of leaf tips and chlorosis. Manganese activates some important enzymes involved in chlorophyll formation. Manganese-deficient plants will develop chlorosis between the veins of their leaves. The availability of manganese is partially dependent on soil pH. Zinc participates in chlorophyll formation and also activates many enzymes. Symptoms of zinc deficiency include chlorosis and stunted growth.

5.3.15 Elemental Analysis

Electron beam excitation is used in electron microscopes, SEM and STEM. X-ray beam excitation is used in XRF spectrometers. A detector is used to convert X-ray energy into voltage signals; this information is sent to a pulse processor, which measures the signals and passes them onto an analyser for data display and analysis. The most common detector now is Si (Li) detector cooled to cryogenic temperatures with liquid nitrogen; however, newer systems are often equipped with SDD with Peltier cooling systems. Chemical characterization is done by using SEM by performing energy dispersive EDX. Results showed that there is no heavy metal contamination or toxic elements present in the soil samples. Very little amount of sample was taken and spread on a pivot and kept in JFC-1600 auto fine coater for 15 minutes. Then the sample is carefully taken out and the pivot containing the sample is kept in

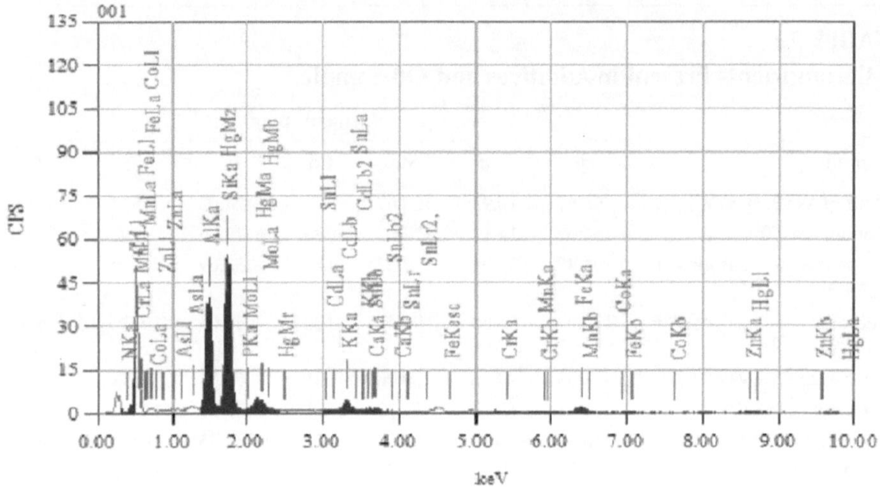

FIGURE 5.12 Composition of different elements of OB sample.

the JSM-6380LA analytical SEM for elemental analysis. Figure 5.12 shows result of EDX analysis and Table 5.7 shows the composition of different elements of OB sample. Analysis shows that there is no heavy metal contamination. Some metals are below detective level whereas some are less than 0.05 At mass%.

Table 5.8 shows the composition of different elements of bottom ash. No heavy metal contamination was found in the analysis. Heavy metals like Cr, Cd, Mo and Hg

TABLE 5.7
ZAF Method Standard Less Quantitative Analysis for OB Sample

Element	(keV)	At Mass%	Error%	Compound Mass%	Cation K
N K	0.392	37.96	2.77	56.35	40.6647
Al K	1.486	19.26	0.48	14.84	19.3404
Si K	1.739	33.75	0.62	24.99	29.9603
K K	3.312	3.27	0.89	1.74	3.7558
Ca K	3.690	0.57	0.92	0.29	0.6912
Cr K	5.411	0.03	1.42	0.01	0.0335
Mn K	5.894	0.03	1.74	0.01	0.0372
Fe K	6.398	3.84	1.79	1.43	4.2114
Co K	6.924	0.02	2.23	0.01	0.0229
Zn K	8.630	0.76	4.69	0.24	0.8028
As K	–	–	–	–	–
Mo L	–	–	–	–	–
Cd L	3.132	0.07	1.88	0.01	0.0609
Sn L	3.442	0.45	2.44	0.08	0.4189
Hg M	–	–	–	–	–
Total		100.00		100.00	

TABLE 5.8
ZAF Method Standard Less Quantitative Analysis for Sewage Sludge

Element	(keV)	At Mass%	Error%	Compound Mass%	Cation K
N K	0.392	54.76	2.89	73.93	62.0976
Mg K	1.253	1.88	0.80	1.46	1.1416
Al K	1.486	9.71	0.74	6.80	6.9182
Si K	1.739	12.71	0.80	8.56	9.5944
S K	2.307	5.96	0.66	3.52	5.6859
K K	3.312	0.46	1.11	0.22	0.4675
Ca K	3.690	4.79	1.16	2.26	5.1599
Cr K	5.411	0.37	1.84	0.14	0.3502
Mn K	–	–	–	–	–
Fe K	6.398	9.09	2.36	3.08	8.3566
Co K	–	–	–	–	–
Zn K	–	–	–	–	–
As K	–	–	–	–	–
Mo L	–	–	–	–	–
Cd L	3.132	0.21	2.33	0.03	0.1677
Sn L	3.442	0.07	3.08	0.01	0.0603
Hg M	–	–	–	–	–
Total		100.00		100.00	

are below detective level. Silicon has the highest At mass% of 36.06 followed by nitrogen of 34.57 At mass%.

Table 5.9 shows the composition of different elements of sewage sludge. Compared to previous results (OB sample and bottom ash), sewage sludge has the higher At mass% of chromium, cadmium and tin but below the permissible level.

Table 5.10 shows the composition of different elements for limestone powder. No heavy metal was found in high concentrations. Sodium and calcium are the two major constituents by At mass%, which together was 86.74 by At mass%. Al, Mn, Co, Fe are less than 1 At mass%.

Table 5.11 shows the composition of different elements of fly ash. No heavy metal concentration was found. Metals like Co, Cd and Cr are less than 0.1 by At mass%. Tin and Zn are less than 1 by at mass %.

Some of the above results were published by the authors (Ram Chandar et al. 2015) and the permission of the publishers to put into this book is highly acknowledged.

5.4 STUDY OF GROWTH OF PLANTS

In order to assess the suitability of mine waste with different additives for growth of plants, a systematic laboratory-scale investigation was carried on. Plants belonging to five different species were chosen for the comparative growth in different samples. The seeds of the plants were bought and sowed individually in OB sample along with

TABLE 5.9
ZAF Method Standard Less Quantitative Analysis for Sewage Sludge

Element	(keV)	At Mass%	Error%	Compound Mass%	Cation K
N K	0.392	54.76	2.89	73.93	62.0976
Mg K	1.253	1.88	0.80	1.46	1.1416
Al K	1.486	9.71	0.74	6.80	6.9182
Si K	1.739	12.71	0.80	8.56	9.5944
S K	2.307	5.96	0.66	3.52	5.6859
K K	3.312	0.46	1.11	0.22	0.4675
Ca K	3.690	4.79	1.16	2.26	5.1599
Cr K	5.411	0.37	1.84	0.14	0.3502
Mn K	–	–	–	–	–
Fe K	6.398	9.09	2.36	3.08	8.3566
Co K	–	–	–	–	–
Zn K	–	–	–	–	–
As K	–	–	–	–	–
Mo L	–	–	–	–	–
Cd L	3.132	0.21	2.33	0.03	0.1677
Sn L	3.442	0.07	3.08	0.01	0.0603
Hg M	–	–	–	–	–
Total		100.00		100.00	

TABLE 5.10
ZAF Method Standard Less Quantitative Analysis for Limestone Powder

Element	(keV)	At Mass%	Error%	Compound Mass%	Cation K
N K	0.392	48.24	2.76	72.78	28.0309
Mg K	1.253	0.35	0.34	0.31	0.2468
Al K	1.486	0.85	0.30	0.67	0.7329
Si K	1.739	5.40	0.28	4.06	5.6731
S K	–	–	–	–	–
Ca K	3.690	38.50	0.41	20.30	57.3018
Cr K	–	–	–	–	–
Mn K	5.894	0.03	0.90	0.01	0.0335
Fe K	6.398	0.67	0.92	0.25	0.7711
Co K	6.924	0.01	1.14	0.00	0.0115
Zn K	8.630	1.27	2.35	0.41	1.4533
As K	–	–	–	–	–
Mo L	–	–	–	–	–
Cd L	–	–	–	–	–
Sn L	3.442	3.64	1.07	0.65	4.1619
Hg M	–	–	–	–	–
Total		100.00		100.00	

TABLE 5.11
ZAF Method Standard Less Quantitative Analysis for Fly Ash

Element	(keV)	At Mass%	Error%	Compound Mass%	Cation K
N K	0.392	37.96	2.77	56.35	40.6647
Al K	1.486	19.26	0.48	14.84	19.3404
Si K	1.739	33.75	0.62	24.99	29.9603
K K	3.312	3.27	0.89	1.74	3.7558
Ca K	3.690	0.57	0.92	0.29	0.6912
Cr K	5.411	0.03	1.42	0.01	0.0335
Mn K	5.894	0.03	1.74	0.01	0.0372
Fe K	6.398	3.84	1.79	1.43	4.2114
Co K	6.924	0.02	2.23	0.01	0.0229
Zn K	8.630	0.76	4.69	0.24	0.8028
As K	–	–	–	–	–
Mo L	–	–	–	–	–
Cd L	3.132	0.07	1.88	0.01	0.0609
Sn L	3.442	0.45	2.44	0.08	0.4189
Hg M	–	–	–	–	–
Total		100.00		100.00	

OB mixed with different additives. Seeds were also sown in native soil to make it as a reference. They were watered daily and weekly growth of the plants was measured using a graduated scale.

Transparent plastic covers were used as pots. Each mix proportion is filled in five covers; each seed belonging to particular specie was sown to a depth of 1–2 inches from the top of the soil. The seeds were sown in such a way that each mix proportion should have all the plant species in isolation. The seeds of the plants chosen are *Cassia siamea, Dendrocalamus strictus, Gliricidia sepium, Phyllanthus emblica,* Peltophorum (Figures 5.13 to 5.17).

- *Cassia siamea*, common name kassod tree, is a fast growing tree that can withstand strong wind and grow well in exposed areas.
- *Dendrocalamus strictus*, also known as Male Bamboo, is a tropical clumping species native to Southeast Asia. It is extensively used as a raw material for paper pulp and has edible shoots.
- *Gliricidia sepium*, often simply referred to as Gliricidia, is a medium-sized leguminous tree belonging to the family Fabaceae. It is considered as the second most important multi-purpose legume tree.
- *Phyllanthus emblica* (syn. Emblica officinalis), the Indian gooseberry (Amla), is a deciduous tree of the family Phyllanthaceae. It is known for its edible fruit of the same name.
- Peltophorum is a genus of between 5–15 species of flowering plants in the family Fabaceae, subfamily Caesalpinioideae. The genus is native to certain tropical regions across the world. The species are medium-sized to large trees growing up to 15–25 m (rarely 50 m) tall.

FIGURE 5.13 Seeds of *Cassia siamea* plant.

FIGURE 5.14 Seeds of *Dendrocalamus strictus* plant.

FIGURE 5.15 Seeds of *Gliricidia sepium* plant.

FIGURE 5.16 Seeds of Peltophorum plant.

FIGURE 5.17 Seeds of *Phyllanthus emblica* plant.

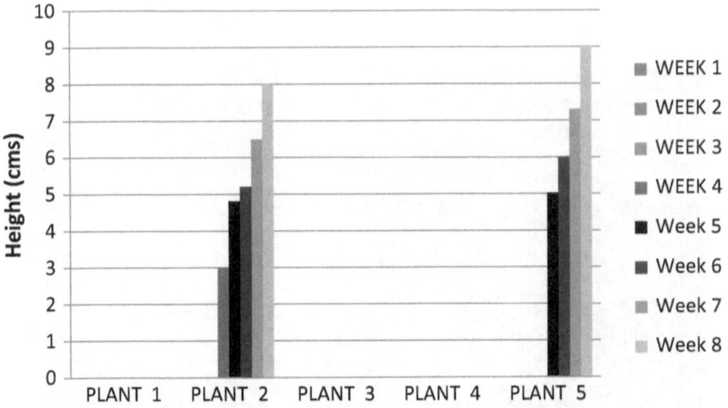

FIGURE 5.18 Growth comparisons of different plant species in fly ash 25% V/V.

The comparative study of growth of the plants in different additives is shown in the following figures.

Figure 5.18 shows growth comparisons of different plant species in OB sample with fly ash 25% V/V. Out of five species only two species have grown in this mix proportion. Plant 2 started growing after 3rd week whereas plant 5 started growing after 4th week. Figure 5.19 shows the growth of Peltophorum plant in fly ash mixed with OB sample in 25% V/V proportion after 8 weeks from sowing of seeds.

Figure 5.20 shows growth comparisons of different plant species in OB sample with fly ash 50% V/V. It is clear that only two species of plants have grown out of five. The rate of growth of plant 2 is being decreased. Plant 4 has started growing after 3rd week and the growth is almost linear. Figure 5.21 shows the growth of

FIGURE 5.19 Growth of Peltophorum plant in fly ash mixed with OB sample in 25% V/V proportion.

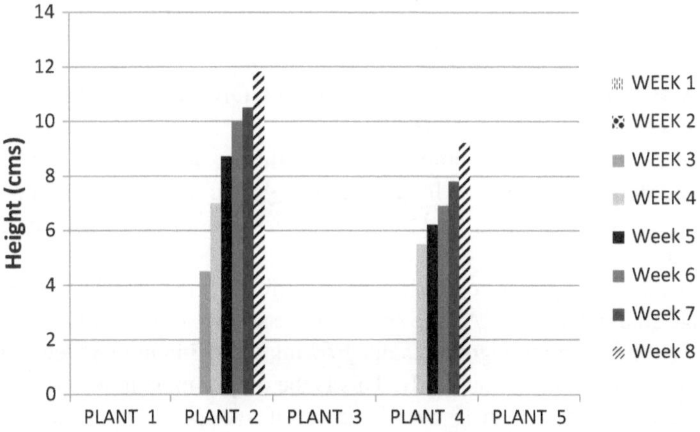

FIGURE 5.20 Growth comparisons of different plant species in fly ash 50% V/V.

FIGURE 5.21 Growth of *Dendrocalamus strictus* plant in fly ash mixed with OB sample in 50% V/V proportion.

Dendrocalamus strictus plant in fly ash mixed with OB sample in 50% V/V proportion after 8 weeks from sowing of seeds.

Figure 5.22 shows growth comparisons of different plant species in OB sample with sewage sludge 25 % V/V. All species have grown in this sample. Plant 3 had shown substantial growth compared to other species. Plant 2 and plant 5 have started growing from week 3.

Figure 5.23 shows growth comparisons of different plant species in OB sample with sewage sludge 50% V/V. All the species have shown good results and plants1, 2 and 3 have shown extremely good results growing up to 34 cm in 8 weeks, which is the maximum height during this study. This is the only sample in which all the plant species have grown more than 15 cm. Three out of five species have started growing from week 1 whereas other two started growing from week 2. This may be due to the high nutrient content present in sewage sludge. Sewage sludge has the highest % of

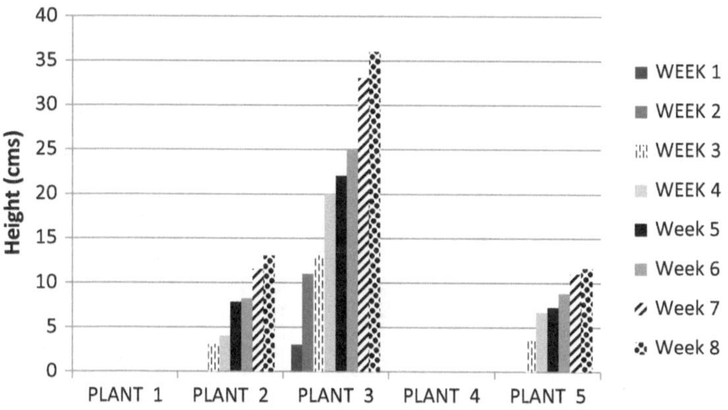

FIGURE 5.22 Growth comparisons of different plant species in sewage sludge 25% V/V.

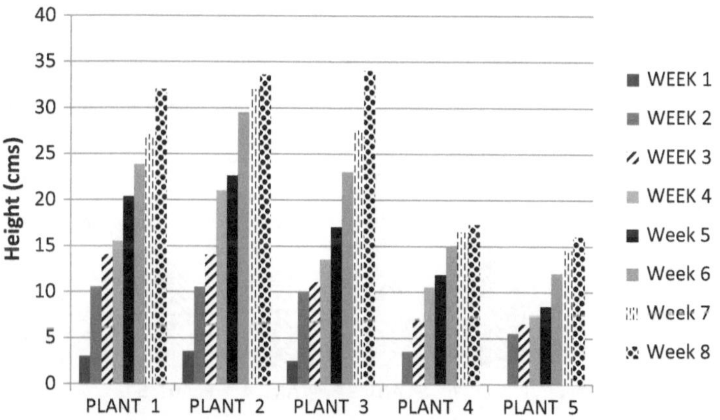

FIGURE 5.23 Growth comparisons of different plant species in sewage sludge 50% V/V.

organic carbon, organic matter and clay content which are supporting the growth of plants.

Figure 5.24 shows growth comparisons of different plant species in OB sample with bottom ash 25% V/V. Out of five species, all species have grown except for plant 1. The rate of growth had been decreasing for plant 3, plant 4 and plant 5 whereas it was almost constant for plant 2.

Figure 5.25 shows growth comparisons of different plant species in sample with bottom ash 50% V/V. Out of five species, all species have grown except for plant 1. Plant 2 and plant 3 have shown significant growth. There is a little restriction of growth during week 5 and week 6 for plant 5.

Figure 5.26 shows the growth comparisons of different plant species in native soil. All the species have shown healthy results showing growth from the very 1st week. Plant 2 and plant 3 have grown more than 25 cm in 6 weeks and all the other species have grown more than 15 cm in 6 weeks.

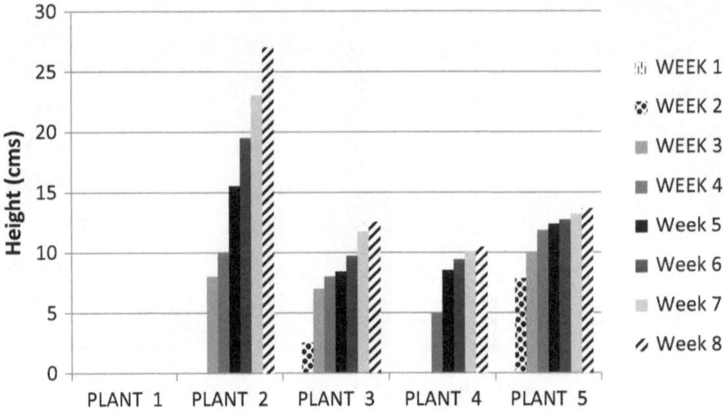

FIGURE 5.24 Growth comparisons of different plant species in bottom ash 25% V/V.

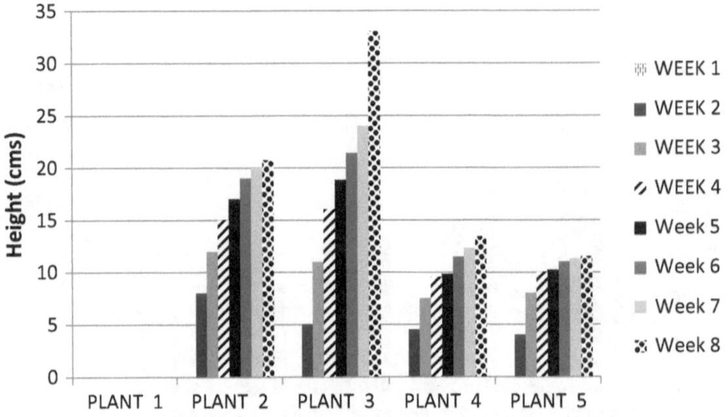

FIGURE 5.25 Growth comparisons of different plant species in bottom ash 50% V/V.

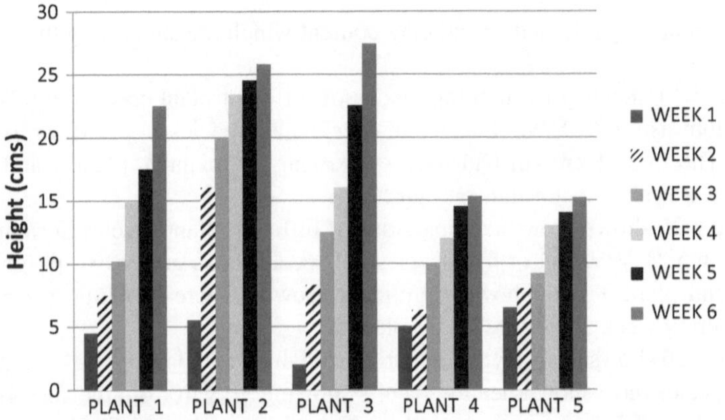

FIGURE 5.26 Growth comparisons of different plant species in native soil.

where,

Plant 1: *Cassia siamea*
Plant 2: *Dendrocalamus strictus*
Plant 3: *Gliricidia sepium*
Plant 4: *Phyllanthus emblica*
Plant 5: Peltophorum

5.5 SUMMARY

Cassia siamea and *Dendrocalamus strictus* have shown better growth (23.8 cm and 29.5 cm, respectively) in sewage sludge mixed with OB sample in 50% V/V proportion than native soil (22.5 cm and 25.8 cm, respectively). All the remaining species have shown better growth results in native soil. Though bottom ash didn't show good experimental results as sewage sludge, the growth of plants is almost same as sewage sludge mixed with OB sample. Peltophorum has grown to a little more height (12.7 cm) in 6 weeks in bottom ash mixed with OB sample in 25% V/V proportion than sewage sludge mixed with OB sample in 25% V/V and 50% V/V proportion (8.8 cm and 12 cm, respectively). Except for *Cassia siamea*, remaining all the species have grown in both the proportions of bottom ash additive whereas only three species have grown in sewage sludge mixed with OB sample in 25% V/V and four species have grown in sewage sludge mixed with OB sample in 50% V/V. But overall growth rate is high in sewage sludge than in bottom ash additive when height is taken into consideration. Though a total of three species have grown in both proportions of fly ash, they are weak and growth is very less compared to other additives.

REFERENCES

Adams, L.M., Capp, J.P. and Gilmore, D.W. 1972. Coal mine spoil and ref use bank reclamation with power plant fly ash. In *Proc. 3rd Mineral Waste Utilization Symposium*, Chicago-USA, March, 1972, 105–111.

Andrews, S.S., Karlen, D.L. and Mitchell, J.P. 2002. A comparison of soil quality indexing methods for vegetable production systems in Northern California. *Agriculture, Ecosystems & Environment*, 90, 25–45.

Adriano, D.C., ed. 1986. *Elements in the Terrestrial Environment*. Springer Verlag, New York, 145–160.

Agarwal, M., Singh, J., Jha, A.K. and Singh, J.S. 1993. *Coal-based Environmental Problems in a Low Rainfall Tropical Region*. Lewis Publishers, Boca Raton, 27–57.

Akala, V.A. and Lal, R. 2001. Soil organic pools and sequestration rates in reclaimed mine soils in Ohio. *Journal of Environmental Quality*, 30, 2090–2104.

Akers, J.D. and Muter, B.R. 1974. Gob pile stabilization and reclamation. *Proceedings of the Fourth Mineral Waste Utilization Symposium*, Chicago, IL, March, 1974, 229–239.

Anne Naeth, M. and Sarah, R. Wilkinson. 2013. Can we build better compost? Use of waste drywall to enhance plant growth on reclamation sites. *Journal of Environmental Management*, 129, 503–509.

Annual Report 2012, *Ministry of Coal*, Government of India.

Anon. 2006. *Dirty Metal, Mining Communities and Environment, Earthworks*, Oxfam America, Washington, DC, 4.

Arshad, M.A. and Coen, G.M. 1992. Characterization of soil quality: Physical and chemical criteria. *American Journal of Alternative Agriculture*, 7, 25–30.

Barnhisel, R.I. 1988. *Correction of Physical Limitation to Reclamation. Reclamation of Surface-Mined Lands*, Vol. 1, Chapter 5. Lloyd R. Hossner, ed. CRC Press, Inc., Boca Raton, FL, 192–211.

Bell, L.C. 2002. *Physical Limitations the Restoration and Management of Derelict Land, Modern Approaches*. World Scientific, New Jersey, 38–49.

Biggest Coal mines in the world, Mining-technology.com (1-06-2014).

Blaylock, M.J. and Huang, J.W. (2000) Phytoremediation of toxic metals: Using plants to clean up the environment. In I. Raskin and B.D. Ensley, eds., *Phytoextraction of Metal*. John Wiley and Sons, Toronto, 303.

Bose, L.K. 2003. Sustainable development of Indian Coal Industry. In A. K. Ghose and L.K. Bose, eds., *Mining in the 21st Century –Quo Vadis?* Oxford and IBH, New Delhi, 191–205.

Bowen, Brian H. and Irwin, Marty W. 2008. *Coal Characteristics- Basic Facts File #8*. Indiana Center for Coal Technology Research.

Bradshaw A.D. 1987. The reclamation of derelict land and the ecology of eco systems. In W.R. Jordan III, M.E. Gilpin and J.D. Aber, eds., *Restoration Ecology: A Synthetic Approach to Ecological Research*, Cambridge University Press, Cambridge, 53–74.

Bradshaw A.D. 1997. Restoration of mined lands- using natural processes. *Ecological Engineering*, 8, 255–269.

Capp, J.P. 1978. Power plant fly ash utilization for land reclamation in the Eastern United States. Reclamation of drastically disturbed lands. *Soil Science Society of America Journal*, 19, 339–353.

Carlson, C.L. and Adriano, D.C. 1993. Environmental impacts of coal combustion residues. *Journal of Environmental Quality* 22, 227–247.

Carr, Phillipa. and Filer, Colin. 2003. Plans to close the OK TEDI mine in Papua New Guinea. Australian National University, 4–6.

Conrad, P.W. 2002. An evaluation of the impact of spoil handling methods on the physical properties of a replaced growing medium on tree survival. Ph.D. Dissertation. University of Kentucky, 262.

Coppin, N.J. and Bradshaw, A.D. 1982. The establishment of vegetation in quarries and open-pit non-metal mines. *Mining Journal Books, London*, 112.

Dickinson, N.M. 2002, Soil degradation and nutrients. In A. D. Bradshaw and Ming H. Wong, eds. *The Restoration and Management of Derelict Land, Modern Approaches*, World Scientific, New Jersey, 50–65.

Donahue, R.L., Miller, R.W. and Shickluna, J.C. 1990. *Soils: An Introduction to Soils and Plant Growth* (5th ed.). Prentice-Hall, 234.

Doran J.W. and Sarrantonio, L.M.A. 1996. Soil health and sustainability. *Advances in Agronomy*, 56, 1–54.

Doran, J.W. and Parkin, T.B. 1996. Quantitative indicators of soil quality: A minimum data set. In J.W. Doran and A.J. Jones, eds. *Methods for Assessing Soil Quality*. SSSA special public. 49. SSSA, Madison, WI, 25–37.

Doran, J.W., Sarrantonio, M. and Liebig, M.A. 1996. Soil health and sustainability. *Advances in Agronomy*, 56, 1–54.

Evans K.G. 2000. Methods for assessing mine site rehabilitation design for erosion impact. *Australian Journal of Soil Research*, 38(2), 231–248.

Foth, H.D. and Ellis, B.G. 1997. *Soil Fertility* (2nd ed.). Lewis CRC Press LLC, Boca Raton, 290.

Gagnon, B., Simard, R.R., Robitaille, R., Goulet, M. and Rioux, R. 1997. Effect of composts and inorganic fertilizers on spring wheat growth and N uptake. *Canadian Journal of Soil Science*, 77, 487–495.

Gardea-Torresdey, J.L., Polette, L., Arteaga, S., Tiemann, K.J., Bibb, J. and Gonzalez, J.H. 1996. Determination of the content of hazardous heavy metals on Larrea tridentata grown around a contaminated area. In L.R. Erickson, D.L. Tillison, S.C. Grant, and J.P. McDonald, eds., *Proceedings of the Eleventh Annual EPA Conf. On Hazardous Waste Research*, NM, 660.

Ghose, M.K. 2001. Management of topsoil for environmental reclamation of coal mining areas. *Environmental Geology*, 40, 1405–1410.

Ghose, M.K. 2005. Soil conversation for rehabilitation and revegetation of mine degraded land. *TIDEE-TERI Information Digests on Energy and Environment* 4(2), 137–150.

Ghosh, A.B., Bajaj, J.C. Hassan, R. and Singh, D. 1983. *Soil and Water Testing Methods- A Laboratory Manual*, IARI, New Delhi, 31–36.

Gitt, M.J. and Dollhopf, D.J. 1991. Coal waste reclamation using automated weathering to predict lime requirement. *Journal Environmental Quality* 20, 285–288.

Gould, A.B., Hendrix, J.W. and Ferriss, R.S. 1996. Relationship of mycorrhizal activity to time following reclamation of surface mine land in western Kentucky. I Propagule and spore population densities. *Canadian Journal Botany* 74, 247–261.

Graves, D.H., Warner, R.C., Wells, L.G., Pelkki, M., Ringe, J.M., Stringer, J., Dinger, J.S., Wunsch, D.R. and Sweigard, R.J. 1995. An interdisciplinary approach to establish and evaluate experimental reclamation of surface mine soil with high value tree species. Unpublished University of Kentucky Project Proposal, 16.

Guerrero, C., Gomez, I. Moral, R. Mataix-Solera, J. Mataix-Beneyto, J. and Hernandez, T. 2001. Reclamation of a burned forest soil with municipal waste compost: Macronutrient dynamic and improved vegetation cover recovery. *Bioresource Technology* 76, 221–226.

Gupta, R.K. and Abrol, I.P. 1990. Salt-affected soils: Their reclamation and management for crop production. In R. Lal and B. A. Stewart, eds., *Advances in Soil Science, Vol. 11*. Springer, New York, 223–288. doi: 10.1007/978-1-4612-3322-0_7.

Hausenbuiller, R.L. 1985. *Soil Science: Principles and Practices* (3rd ed.) Wm. C. Brown, Iowa, 610.

Hobbs R.J. and Norton D.A. 1996: Towards a conceptual framework for restoration ecology. *Restoration Ecology* 4, 93–110.

IBM. 2000. Reclamation/restoration techniques and strategy for mined out areas (Bulletin no. 37). Indian Bureau of Mines, Napur, 122.

IEA and BP Data Published in 2013, World Coal Association. www.worldcoal.org, 2nd July 2014).

Indorante S.J., Jansen I.J. and Boast C.W. 1981. Surface mining and reclamation: Initial changes in soil character. *Journal of Soil and Water Conservation*, 36, 347–351.

Jackson, L.L., Lopoukhine, N. and Hillyard, D. 1995. Ecological restoration: A definition and comments. *Restoration Ecology* 3(2): 71–75.

Jenness, N. 2001. Mine reclamation using biosolids. U.S. Environmental Protection Agency Office of Solid Waste and Emergency Response Technology Innovation Office.

Johnson, D.B. and Williamson, J.C. 1994. Conservation of mineral nitrogen in restored soils at opencast mines sites: I. Result from field studies of nitrogen transformations following restoration. *European Journal of Soil Science* 45, 311–317.

Jurinak, J.J. and Suarez, D.L. 1990. The chemistry of salt-affected soils and waters. In K.K. Tanji (ed.) *Agricultural salinity assessment and management*. Am. Soc. Civil Eng. Manuals Reports on Eng. Practice no. 7 1. ASCE, New York, 42–63.

Karlen, D.L., Eash, N.S. and Unger, P.W. 1992. Soil and crop management effects on soil quality indicators. *American Journal of Alternative Agriculture*, 7, 48–55.

Karlen, D.L., Maubach, M.J., Doran, J.W., Cline, R.G., Harris, R.F. and Schuman, E. 1997. Soil quality, a concept, definition and framework for evaluation. *Soil Science Society of America Journal* 61, 4–10.

Kavamura, V.N. and Espostio, E. 2010. Biotechnological strategies applied to the decontamination of soil polluted with heavy metals. *Biotechnology Advances*, 28, 61–69.

Kostadinov, S. and Marković, S. 1996. Soil erosion and effects of erosion control works in the torrential watersheds in South—East Serbia. In D.E. Walling and B.W. Webb, eds., *IAHS Publication no 236 ISSN 0144-7815 "Erosion and Sediment Yield: Global and Regional Perspectives"*. IAHS Press Walingford, England, 321–332.

Kozlowski, T.T. 1999. Soil compaction and growth of woody plants. *Scandinavian Journal of Forest Research* 14(6) 596–619.

Kumar, N. 2007, Soil quality standards (SQS) for bio-reclamation of coal overburden dumps: ISO-14000 requirements. *Journal of Industrial Pollution Control*, 23 (1), 19–23.

Kundu, N.K. and Ghosh, M.K. 1994. Studies on topsoil of opencast coal mine. *Environmental Conservation*, 21, 126–132.

Lado, M., Paz, A. and Ben-Hur, M. 2004. Organic matter and aggregates size interaction in saturated hydraulic conductivity. *Soil Science Society of America Journal* 68, 234–242.

Lal, R. and Pierce, F.J. 1991. The vanishing resource. In R. Lal and F.J. Pierce, ed. *Soil Management for Sustainability*. Soil and Water Conserv. Soc., Ankeny, IA, 1–5.

Lindsay, W.L. and Norvell, W.A. 1978. Development of a DTPA soil test for zinc, iron, manganese, and copper. *Soil Science Society of America Journal* 42: 421–428.

Lone, M.I., He, Z. L., Stoffella, P. J. and Yang, X. 2008. Phytoremediation of heavy metal polluted soils and water: Progress and perspectives. *Journal of Zhejiang University Science B*, 9(3), 210–220.

Maiti, S.K. 1994. Some experimental studies on ecological aspects of reclamation in Jharia coalfield. Ph.D. thesis submitted to Indian School of Mines, Dhanbad.

Maiti, S.K., Karmakar, N.C., Sinha, I.N. 2002. Studies into some physical parameters aiding biological reclamation of mine spoil dump – A case study from Jharia coal field. *Indian Journal of Mining & Engineering* 41, 20–23.

Maiti, S.K. and Ghose, M.K. 2005. Ecological restoration of acidic coal mine overburden dumps- an Indian case study. *Land Contamination and Reclamation* 13(4), 361–369.

Maiti, S.K. and Singh, G. 2001. Ecosystem recovery in an afforested coalmine overburden dump-problems and recommendations. In *Proceedings on Reclamation and Rehabilitation of Mined Out Areas*, organized by SGAT, Bhubaneswar, India, 16–17th Feb. 2001, 125–129.

Mamo, M., Molina, J.A.E., Rosen, C.J. and Halbach, T.R. 1999. Nitrogen and carbon mineralization in soil amended with municipal solid waste compost. *Canadian Journal of Soil Science* 79, 535–542.

Marschner, H. 1995. *Mineral Nutrition of Higher Plants*. Academic Press, London.

Martínez, F., Cuevas, G., Calvo, R. and Walter, I. 2003. Biowaste effects on soil and native plants in a semiarid ecosystem. *Journal of Environmental Quality*, 32, 472–479.

McBride, M.C. 1994, *Environmental Chemistry of Soils*. Oxford University Press, Oxford.

Meagher, R.B. 2000. Phytoremediation of toxic elemental and organic pollutants. *Current Opinion in Plant Biology*, 3: 153–162.

Munshower, F.F. 1994, *Practical Handbook of Disturbed Land Revegetation*. Lewis Publishers, Boca Raton, Florida.

Nova Scotia Environment. 2009. Guide for surface coal mine reclamation plans. http://novascotia.ca/nse/ea/ (2nd July 2014).

Pederson, T.A., Rogowski, A.S. and Pennock, R. 1988. Physical characteristics of some mine spoils. *Soil Science Society American Journal*, 44, 131–140.

Pierce, F.J., Larson, W.E., Dowdy, R.H. and Graham, W.A.P. 1983. Productivity of soils: Assessing long term changes due to erosion. *Journal of Soil and Water Conservation* 38(1), 39–44.

Punmia, B.C., Jain, A.K. and Jain, A.K. 2005. *Soil Mechanics and Foundations*. Laxmi Publications, New Delhi, 98–101.

Ram Chandar, K. Chaitanya, V. and Raghunandan, M.E. 2015. Experimental Study for the Assessment of Suitability for Vegetation Growth on Coal Over Burden. *International Journal of Mining and Mineral Engineering*, 6(3), 218–233.

Rankin, W.J. 2011. *Minerals, Metals and Sustainability: Meeting Future Material Needs*. CSIRO Pub., Collingwood, Vic.

Richards, L.A., ed., 1954. Diagnosis and improvement of saline and alkali soils. U. S. Salinity Laboratory. Agriculture Handbook No. 60. U. S. Dept. of Agri., Washington, D.C.

Ros, M., Garcia, C. and Hernandez, T. 2001. The use of urban organic wastes in the control of erosion in a semiarid Mediterranean soil. *Soil Use Management* 17, 292–293.

Sahu, H.B. 2011. Land degradation due to mining in India and its mitigation measures. In *Proceedings of Second International Conference on Environmental Science and Technology*, 1–4.

Sanchez, C.A. 2007. Phosphorus. In A.V. Barker and D.J. Pilbeam, eds., *Handbook of Plant Nutrition*. CRC Press, Boca Raton, FL.

Sanyal, B. 2006. Coal to stay key source for nation's energy needs. *Business Line*, November 22, 2006.

Serra-Wittling, C., Houot, S. and Barriuso, E. 1996. Modification of soil water retention and biological properties by municipal solid waste compost. *Compost Science & Utilization*, 4, 44–52.

Sheoran, V., Sheoran, A.S. and Poonia, P. 2010. Soil reclamation of abandoned mine land by revegetation: A review. *International Journal of Soil, Sediment and Water Documenting the Cutting Edge of Environmental Stewardship*, 3(2), 1–4.

Singh, A.N. and Singh A.N. 2006. Experiments on ecological restoration of coal mine spoil using native trees in a dry tropical environment, India: A synthesis. *New Forests*, 31, 25–39.

Singh, R.S., Chaulya, S.K., Tewary, B.K. and Dhar, B.B. 1996. Restoration of coal-mine overburden dump – a case study. *Coal International*, 33(7), 331A–331A

Singh, R.S., Tewary, B.K. and Dhar, B.B. 1994. Effect of surface mining on plant biomass and productivity in a part of Dhanbad coal field areas. In S.P. Banerjee, ed., *Second National Seminar on Minerals and Ecology*. Oxford & IBH Pub, New Delhi, 103–109.

Society for Ecological Restoration International Science & Policy Working Group 2004. The SER International Primer on Ecological Restoration. www.ser.org, 5th Dec., 2013.

Stewart, B.R. and Daniels, W.L. 1995. The impacts of coal refuse/fly ash bulk blends on water quality and plant growth. In G.E. Schuman and G.F. Vance, eds., *Proc., 12th Ann. Meet., Amer. Soc. Surf. Mining and Rec., ASMR*. Lexington, KY, 105–116.

Taiz, L. and Zeiger, E. 1991. *Plant Physiology*. Sinauer Associates, Sunderland, p. 690.

Tester, C.F. 1990. Organic amendments effects on physical and chemical properties of a sandy soil. *Soil Science Society of America Journal* 54, 827–831.

Tisdale, S. Nelson, W.L. Beaton, D. and Havlin, J.L. 1993. *Soil Fertility and Fertilizers*. (5th ed.). MacMillan Publishing Company. New York USA.

Trivedi, R.K., Goel, P.K. and Trisal, C.L. 1987, *Practical Methods in Ecology and Environmental Science*. Environmental Publications, Karad, Maharashtra.

USDA (U.S. Department of Agriculture). 1951. Handbook #18. *Soil Survey Manual*, 503.

USDA (U.S. Department of Agriculture). 1964. *Soil Conversation Service Engineering Handbook*, Section 15: Irrigation, Chapter 1: Soil-plant-water relationships.

Van Andel, J. and Aronson, J., eds. 2006. *Restoration Ecology*. Blackwell, Oxford.

Van Bruggen, A.H.C. and Semenov, A.M. 2000. In search of biological indicators for soil health and disease suppression. *Applied Soil Ecology*, 15, 13–24.

Wali, M.K. 1987. The structure dynamics and rehabilitation of drastically disturbed ecosystems. In *Perspectives in Environmental Management*. Oxford Publications, New Delhi, 163–183.

Walker, L.R., Walker, J. and Hobbs, R.J., eds. 2007. *Linking Restoration and Ecological Succession*. Springer, New York.

Wong, M.H. 2003. Ecological restoration of mine degraded soils, with emphasis on metal contaminated soils. *Chemosphere* 50, 775–780.

6 Utilization of Iron Ore Tailings in Bricks

Shubhananda Rao P. and Ram Chandar Karra

CONTENTS

DOI: 10.1201/9781003268499-6

6.1 INTRODUCTION

The current trend all over the world is to utilize the treated industrial by-products as raw material in the construction, which gives an eco-friendly solution to waste disposal. To achieve this objective, intensive efforts are underway for effective utilization of industrial by-products particularly from mining and mineral industries. The role of additives will enhance the physico-mechanical properties of the materials, adhering to standard recommendation to the construction.

The waste produced during processing of metallic minerals is called tailings. The iron ore tailings (IOT) have many adverse impacts on the environment like pollution of groundwater, soil erosion, loss of biodiversity, soil contamination, infertility of soil, acid mine drainage, etc. The use of such tailings in the construction material has double benefit of reduction in the cost of the product and safe disposal of tailings safeguarding the environment. With IOT bricks, tailings storage and disposal can be avoided. Land requirement for tailings dam can be avoided. Lot of environmental problems in handling the IOTs can be reduced. One such usage can be in manufacture of building bricks and other value-added products without burning of solid fuel, which helps in the reduction of CO_2 emission and saves natural resources like sand and clay. Some researchers have studied the use of different industrial waste in making bricks, which is summarized in Chapter 2.

The bricks made of IOT will have higher density because of iron content, which needs to be reduced without hampering the required compressive strength and other physico-chemical properties.

Energy consumption has been increasing rapidly worldwide due to strong growth in population and industrialization. Buildings are responsible for about

40% of total energy consumption and one third of greenhouse emissions in the world. Pure thermal performance of building envelope structure is the main cause of energy consumption in buildings. In a typical house, external walls account for approximately 30% of total heat loss which requires an effort to improve heat insulation performance of each wall component (Arici et al. 2016). So, perlite bricks can be used in the walls to reduce the heat transfer through bricks and the use of perlite in the bricks is an initiative to green building concept having long-term benefit of sustainability.

6.1.1 Energy Consumption in Production of Bricks

The embodied energy of the brick is the sum of all the energy required to produce the bricks in all processes from quarrying of clay to firing of bricks. Energy inputs usually entail greenhouse gas (GHG) emissions. It decides how much brick production contributes to global warming. The embodied energy is measured in MJ/kg. A non-fired brick has the advantage of lower embodied energy and in turn lower GHG emission.

Toledo et al. (2004) analysed the gas release, crystalline structure and ceramic properties formed during firing of clay raw materials and extruded bricks. CO, CO_2 and NO_2 and methane emission were measured during firing cycle and found CO_2 emitted from the powder was 8600 PPM and from the extruded samples was 6500 PPM, CO emission was found to be 1100 PPM from the powder and 800 PPM from the brick and with some minor emission of NO_2 and CH_4.

Michael et al. (2009) surveyed on the energy consumption and the GHG emission during the clay brick production in the United States. Embodied energy for common clay was about 9.3 MJ/brick. The GHG emission per common clay brick fired using fossil fuel was around 0.6 kg of CO_2 to the atmosphere, whereas one common concrete brick emits 0.3 kg of CO_2.

Gonzalez et al. (2011) has highlighted about the environmental contamination represented by the enormous emissions of GHG, resulting in unusual climate changes as smog, acid rain and global warming. Hence, recycling the wastes in the bricks production appears to be not only a viable solution to environmental pollution but also an economical option to design green buildings. However, the chronic problem of GHG and energy consumption has not yet been tackled properly as many of the previous research works were mainly focused on recycling the waste traditionally in the bricks.

Li et al. (2019) discussed and concluded about recycling of industrial waste IOT in porous bricks with low thermal conductivity. The results showed that sintering temperature, soaking time and milling time had significant effects on porosity, compressive strength and microstructure of the porous tailing ceramics. Extending soaking time, meanwhile decreasing sintering temperature could effectively improve the compressive strength of sample with similar porosity. The thermal conductivity of the porous tailing bricks could reach the lowest value of 0.032 W/mk with the porosity of 89%. The thermal conductivity of samples with varied porosity is comparable with the calculated values deducted from the universal model, which makes a good description of the relationship between porosity and thermal conductivity.

6.1.2 Perlite as Density Controller

Any structure designed intelligently and responsibly aspires to be as light as possible. The function is to support live loads. The dead loads of the structure itself are a necessary evil. The smaller the ratio between a structure's dead load and the supported live load, the lighter the structures are. From an ecological, social and cultural perspective, lightweight structures have never been more contemporary and necessary than today. The use of bricks as void filler in the framed structure has given rise to the concept of reducing the dead load in the buildings. Hence, a lightweight additive needs to be added to reduce the density. Perlite is one such additive which has very low density of around 60 kg/m^3, which can be used as additive to reduce the weight of bricks.

Perlite is an amorphous volcanic rock that has relatively high water content, typically formed by the hydration of obsidian. It occurs naturally and has the unusual property of greatly expanding when heated sufficiently. It is an industrial mineral and a commercial product useful for its low density after processing. Perlite softens when it reaches temperature of 850–900°C. Water trapped in the structure of the material vapourizes and escapes, and this causes the expansion of the material to 7–16 times of its original volume.

Unexpanded perlite has a bulk density of around 1100 kg/m^3, while typical expanded perlite (EP) has a bulk density of about 30–150 kg/m^3 (Samar and Saxena 2016). Because of light weight and better insulating characterization of perlite, it is used in lightweight concrete, loose fill masonry insulation, chimney linings, etc. The use of perlite in bricks may act as density controller and will impact many other advantages like heat insulation, light weight, acoustic insulation properties, etc.

Bulut (2010) stated that perlite has chemical inertness, fire resistance and high absorption of sound. All these properties make perlite a usable material for many applications. The EP can be used in the construction industry and horticulture market and as a filter aid and filler. This lightweight filler is used as insulating cover on the surface of the molten metal to prevent excessive heat loss during delays in pouring, to top of ingots, to reduce piping and decrease lamination, to produce refractory blocks and bricks or simply as fillers and in several important foundry applications.

One of the main problems associated with the production of EP is formation of relatively large quantities of fine fraction (about 5%–10%) with a particle size below 200μm. The resulting lightweight waste due to a large specific surface area and dustiness is very difficult to dispose, especially in dry conditions. EP is fire-, heat- and chemical-resistant material that has porous structure. Porosity is determined as the average ratio between volume of pores and total volume of perlite grains. Porous structure gives perlite volumetric and surface absorption capability. To prevent water pollution and to ensure insulation, water absorption is involuntary because pores filled with water increase heat conductivity. Porous structure gives perlite surface absorption and lightweight property. The unit weight of perlite depends on gradation and expansion. The heat conductivity of dry perlite that has unit weight of 90 kg/m^3 is calculated as 0.04 W/mK at 24°C according to dry unit weight method. Perlite materials have advantage on sound absorption and insulation. Chemically, perlite ore consists of SiO_2, Al_2O_3 and lesser amounts of several metal oxides (sodium, potassium, iron, calcium and magnesium) and therefore can be an attractive addition in many of the building materials like concrete, brick, etc., because of its excellent insulation properties and relatively high compressive strength despite a very low bulk density (Samar and Saxena 2016).

6.1.3 Use of Perlite in Bricks

Demir and Orhan (2004) investigated on the production of construction bricks with perlite addition. The production of light building brick with adding of perlite into the clays was aimed. The study of chemical, mineralogical and thermal analysis was carried out on perlite brick. In order to get comparable results, the materials produced to test were added into the brick clay which is by weight: A: 0% perlite, B: 2.5% and C: 5% EP. Test specimens were produced by vacuum extrusion with a press having cross-section of 75 mm × 40 mm, 100 mm length and 35% perforation volume. The samples were tested by using the standard test methods and compared with the specifications and evaluated. Porosity and water absorption values increased in the perlite addition series. Thermal conductivity values of sample drastically decreased with increase in amount of EP. Compressive strength with addition of 5% perlite is 8.72 MPa compared to bricks without perlite 9.25 MPa is less but it is higher than the standard. As a result, it was concluded that perlite addition shows a behaviour of lightweight and insulation building brick.

Lanzon and Garcia-Ruiz (2008) studied the influence of perlite on the fresh and hardened state of cement mortar. Water absorption, workability, mechanical strength, and sorptivity were reported. The outcomes indicated that water absorption, sorptivity, and mechanical strength induced a negative effect, while water retentivity and workability have improved generally.

Figen et al. (2010) carried out a research study on development of the insulation materials from coal, fly ash, perlite, clay and linseed oil. The results showed that compressive-tensile strength of the insulation material decreased when the high fly ash ratio and high epoxidized linseed oil ratio used in the preparation of the insulation material composition. The compressive and tensile strengths varied from 10.01 to 1.107 MPa and 8.38 to 1.013 MPa, respectively. The minimum thermal conductivity of 0.313 W/mK observed for the sample made with a 60% FA/30% C/10% P ratio and 50% ELO processed at 200°C. It is increased with the decrease of ELO and FA. Results indicate an interesting potential for the coal fly ash recycling and epoxidized linseed oil renewable to produce useful materials.

Ayudhya (2011) presented the results of an experimental study on the residual compressive and splitting tensile strength of autoclaved aerated concrete (AAC) containing perlite and polypropylene (PP) fibre subjected to high temperatures. Cylinder specimens were subjected to various temperature ranges of 100, 200, 400, 800 and 1,000°C. The mixtures were prepared with AAC cementitious materials containing perlite at 15%, 20% and 30% sand replacement. The polypropylene fibre content of 0, 0.5%, 1%, 1.5% and 2% by volume was also added to the mixture. The results showed that the unheated compressive and splitting tensile strength of AACs containing PP fibre were not significantly higher than those containing no PP fibre.

Furthermore, the presence of PP fibre was not more effective for residual compressive strength than splitting tensile strength. The 30% perlite replacement of sand gave the highest strength. Based on the results, it can be concluded that addition of PP fibre did not significantly promote the residual strength of AAC specimens subjected to high temperatures.

Sengul et al. (2011) used perlite instead of fine aggregate with various replacement ratios depending on required strength. Test results indicated the compressive strength and modulus of elasticity decreased with increasing perlite content in the concrete mixture. Moreover, water absorption and sorptivity co-efficient increased with the higher perlite contents. Replacing normal aggregate by EP reduced the thermal conductivity of the mixtures as a result of the porous structure of the perlite.

Benk and Coban (2012) studied the production of lightweight, heat insulating and water-resistant bricks from lightweight aggregate like pumice and EP. The raw pumice was poured into the water and then floated aggregates were selected for the mixture. The floated pumice was dried and crushed with respect to pass through 1 mm sieve. Crushed pumice was mixed with perlite borax and with blend of molasses and hardener. About 40 bricks were prepared from three types of mixtures. Each brick was fabricated by compressing 20 gm of the mixture in 23 mm internal diameter steel mould. The mix design containing 20% EP, the amount of mixture was diminished to 10 gm. Heat treatment applied was 200°C for 2 hours then 650°C for 1 hour and at last specimen were cured at 825°C for 1 hour. Results showed that the hardened molasses bonded bricks should be preferred. When 2.5% borax was not used in the mixture, considerable reduction in tensile strength of bricks was occurred by alteration of replacement ratio of perlite with pumice.

Celik and Durmaz (2012) studied the compressive strength in different cement types and dosages of concretes made by using 60% pumice and 40% perlite. In this study, it was tried to obtain bearing concrete by using the pumice and perlite aggregates in certain proportions. Different cement dosages were used in every mixture poured with Portland cement and Portland composite cement, the 60% pumice and 40% perlite aggregates of which were kept stable. The silica fume was added as much as 10% of cement amount in mixture. Mixtures from 200 to 500 doses were made and poured with every type of cement. The prepared samples left to water cure and their 7 and 28 days compressive strengths were determined. The test results indicated that the mixtures poured with Portland cement has a higher compressive strength than those poured with Portland composite cement. While the concretes poured with Portland composite cement up to 500 doses can be used only insulation, the concretes poured with Portland cement at 450 doses and above show the characteristics of side bearing concrete. The cement dosage increases the compressive strength in lightweight concrete poured with PC32.5.

Chaouki et al. (2013) produced porous fire brick from mixture of clay and recycled refractory waste with the EP addition. Production of porous and lightweight bricks with acceptable flexural strength is accomplished. EP was used as an additive to an earthenware brick to produce the pores. SEM-EDX, XRD and XRF analysis of the raw materials and the elaborated refractory were performed. Mixtures containing perlite were prepared at different proportions (up to 30% %). Apparent porosity at 1600°C was investigated with the bulk density, water absorption, firing shrinkage and flexural strength. Microstructural investigation was carried out by both natural light microscopy and polarized light microscopy. The results obtained showed that the samples tested here maintained their shape without undergoing any deformation up to 1600°C. The use of EP decreased the fired density of the bricks down to 1.55 g/cm^3.

Celik (2014) investigated the reproducibility of brick using EP aggregate as the main raw material. In additional carboxy methyl cellulose (CMC) as a chemical binder, potassium and sodium borate were used as natural binder and coal powder as an additive. Samples were cured at 400°C for 2 hours. The results indicated the unit weight between 520 and 580 kg/m^3 and compressive strength as 23 kg/cm^2. Thermal conductivity fluctuated between 0.09 to 0.123 W/mk. Perlite bricks with Na or K borate averted harmful sunlight and radiations by its higher neutron absorbility.

Mohajer et al. (2014) studied on lightweight refractory insulation panels on the basis of perlite 30% with chemical bonding (H_3PO_4) showed sintering behavior is the key to achieve right balance between the lower densities, which is the basis for good thermal insulation properties, good mechanical properties and durability, saving the energy that escapes from the industrial furnaces.

Because of perlite's outstanding insulating characteristics and light weight, it is widely used as loose-fill insulation in masonry construction. In this application, free-flowing perlite loose-fill masonry insulation is poured into the cavities of concrete block where it completely fills all cores, crevices, mortar areas and ear holes. In addition to providing thermal insulation, perlite enhances fire ratings, reduces noise transmission and it is rot, vermin and termite resistant. Perlite is also ideal for insulating low temperature and cryogenic vessels. When perlite is used as an aggregate in concrete, a lightweight, fire resistant, insulating concrete is produced that is ideal for roof decks and other applications. Perlite can also be used as an aggregate in Portland cement and gypsum plasters for exterior applications and for the fire protection of beams and columns. Other construction applications include under-floor insulation, chimney linings, paint texturing, gypsum boards, ceiling tiles, and roof insulation boards.

Shankarananth and Jaivignesh (2016) presented a parametric experimental study of utilization of glass powder, GGBS, Perlite ore in fly ash brick manufacturing. Glass powder, Perlite ore is taken as constant of 2%, 10% and GGBS is replaced with 10%, 20% and 30% for each proportion that have been calculated. The physical and mechanical properties of fly ash brick are investigated. These bricks were tested for compressive strength, water absorption, efflorescence, density and soundness test after 21 days curing as per Indian Standards. The result showed greater compressive strength than the ordinary first class bricks and it also concludes better water absorption, 1300–1500 kg/m^3 density and good soundness by a clear ringing sound.

Xu et al. (2016) produced a new type of rubber and perlite mortar modified by SBR latex and polyester. Thermal and mechanical properties were studied. The experimental results showed that the amount of rubber and perlite have a major influence in a compressive strength and thermal conductivity. With increasing rubber and perlite dosages, the compressive strength and thermal conductivity decreases. The reduction in compressive strength was 25%–65%, as compared to the control mortar. The reduction of thermal conductivity was 6%–12% and 30%–35% for rubber and perlite, respectively.

Zulkifeli and Saman (2016) evaluated experimentally the effect of fire on the perlite cement mortar. The sand was replaced by perlite with the contents of 10%, 20%, 30% and 40% by volume. The mortars were exposed to different high temperatures of 200°C, 400°C, 700°C, and 1000°C. The compressive strength was reduced with increasing the perlite content, particularly for low temperature exposure, while the performance improved in high temperature.

Arunraja et al. (2017) investigated mechanical properties of lightweight bricks using perlite and lime. This work effectively converts perlite into useful building materials like building bricks and floor interlocks which can effectively reduce the self-weight or dead load and further decreases the problem of fire industries accidents in the society. Rather than the EP going into the landfill or incinerators it can be used as construction materials at a much lower cost after undergoing certain specific processing. This brick is floating in the water. The specimen size was 190 mm × 90 mm × 90 mm. These bricks were made of perlite and lime in various ratios like 70:30, 75:25, 80:20 and 85:15. From the compression testing results, it was found that EP material when effectively mixed with lime gives the 3.3 N/mm^2 compressive strength.

Naveen et al. (2018) studied the effect of compositions of bagasse ash and perlite admixture on thermal stresses and other mechanical properties of clay. The samples were dried and then finally fired in the furnace at 800°C for a final curing. Their fired densities varied between 1790 and 1230 kg/m^3, which correspond to a decrease of 27%, when compared to the density of the brick without admixture. Apparent water absorption values were increased with increase in Bagasse ash, Perlite addition. Properties which include thermal shock resistance, cold crushing strength and porosity were obtained by the appropriate standard test methods. The microstructures and weight loss percentage (%) corresponding to temperature variation of the fired samples were characterized with SEM-EDS and TGA. The results show that the amount of bagasse ash and perlite admixture affects the properties variously; porosity and thermal resistance increases with percentage increase in bagasse ash and perlite, thermal and mechanical properties were also evaluated by ANSYS.

Satakhun et al. (2018) investigated on masonry and plastering mortars made from Portland cement, fly ash, EP, mortar plasticizer and sand. For the masonry mortar, the mixes with high calcium fly ash with or without plasticizer could be used with required properties and reduced cost. The mix of EP gave water retention over the required 70%. The mix containing fly ash and 0.2% plasticizer gave water retention slightly less than 70%. It was recommended that the mix with EP should be used for indoor plastering and the mix with fly ash and plasticizer should be used for outdoor plastering.

6.1.4 Energy Savings in the Production of Bricks Using Perlite

Topcu and Isikdag (2006) investigated on different methods for achieving heat insulation in the buildings. Manufacturing of high heat conductivity resistant construction materials is an important part of these research efforts. High heat resistant brick can be produced by adding perlite into the clay in conventional brick manufacturing. In this investigation perlite of Eskisehir region and clay were collated and fired to form high heat conductivity resistant material. Binding materials such as cement, gypsum, lime, bitumen and clay were used for manufacturing perlite brick. Bricks in standard sizes manufactured at different perlite-clay ratios and unit weight, compressive strength, volume reduction and heat conductivity values were obtained. Compressive strength decreases, heat conductivity resistant and shrinkage of perlite bricks increase as the replacement ratio of perlite increases. Results were examined according to combined properties, and specialties of perlite bricks were determined at various weights. As a result, the best mixture was determined as the one containing 30% perlite.

Jedidi et al. (2015) carried out an experimental study in order to provide more data on the effects of expanded perlite aggregate (EPA) dosage on the compressive strength and thermo physical properties of lightweight concrete at different ages. The first part of this experimental study was devoted to the choice of the proper mixing procedure for expanded perlite concrete (EPC). Thereafter, six sets of cubic specimens and six sets of parallelepiped specimens were prepared at a water-to-cement ratio of 0.70 with varying replacement percentages of sand by EPA ranging from 0% to 80% by volume of sand. Compressive strength, thermal conductivity and thermal diffusivity were determined over curing age. Unit weights for the mixtures prepared varied between 560 and 1510 kg/m^3. Compressive strength was decreased when perlite content was increased. The test results indicated that replacing natural aggregate by EPA increased the thermal resistance of the lightweight concrete and consequently, improved thermal insulation.

Arici et al. (2016) investigated on heat insulation performance of hollow clay bricks filled with perlite. Heat transfer through various hollow bricks, which are widely used in Turkey, is analysed numerically. Calculations are carried out for three scenarios for each type of hollow brick: (i) cavities are filled with air (ii) half of the cavities are filled with perlite while the other half with air (iii) all cavities are filled with perlite. The computed results showed that filling the cavities with perlite inhibits convection and radiation in cavities. Hence, effective thermal conductivity, thus heat loss (or heat gain), reduces significantly for all types of bricks. The enhancement in heat insulation performance can be up to 15.6% and 27.5% for half-perlite and full-perlite cases, respectively, depending on the brick type. It is concluded that since filling the cavities of hollow bricks with perlite provides a high potential of reducing effective conductivity, it should be considered by engineers to build residential or office buildings with low energy consumption.

Georgiev et al. (2017) studied on effect of expanded vermiculite and EP as pore forming additives on the physical properties and thermal conductivity of porous clay bricks. It has been established the effect of the quantity of 0, 3, 5 and 8 mass% of two pore-forming additives as expanded vermiculite and EP on the properties of the fired at 900°C porous ceramic brick samples. The apparent density of the fired ceramic brick samples decreases with the increasing of pore-forming amount due to the increased number of pores created by expanded vermiculite and EP during firing. Increasing the amount of pore-forming additive increases porosity and decreases the thermal conductivity of the samples. For contents of small amounts of pore-forming additive in the ceramic bodies, the changes of physical properties and thermal conductivity are less pronounced. Thermal conductivity of the porous brick sample with

8 mass% vermiculite, produced at 900°C compared to the brick without additive, decreases from 1.1 to 0.8 W/mK. Expanded vermiculite and EP in amounts of 8 mass% could be used as a pore-forming agent in bricks production to improve thermal conductivity of clay bricks and keep acceptable compressive strength.

Bullibabu and Ali (2018) investigated on the reduction of thermal conductivity in clay-based construction materials. Clay is the basic material for brick production; bagasse and perlite are additive elements in clay; bagasse is a sugar industrial waste; and perlite is a siliceous volcanic rock. Bagasse facilitates the porous nature in brick and perlite consists of rich amount of Al_2O_3 and SiO_2, while thermal insulation properties improve in clay-based construction materials. The result shows that an increasing quantity of the bagasse and perlite in the clay mixture significantly decreases the thermal conductivity of clay brick, and a small reduction in compressive strength is observed.

Uluer et al. (2018) mathematically calculated and experimentally investigated on expanded perlite-based heat insulation materials' thermal conductivity. Thermal resistance can be increased by using proper heat insulation materials. Traditional heat insulation materials do not stand all desired properties. Thus, developing new heat insulation materials is very important. In this study, expanded perlite-based heat insulation material was developed as an alternative to the traditional insulation materials. The composition of the developed material was designed and prepared using the theoretical thermal conductivity prediction models. The prepared material was moulded in a rectangular shape panel. Thermal conductivity of panels was measured experimentally and the results were compared with the calculated results. Also, the results showed that the developed panels can be used for heat insulation applications. On the other hand, the closest model to the experimental results is the parallel model whose average deviation is 4.22% while the farthest model is the Cheng and Vachon model whose average deviation is 12.43%. It is observed that parallel and series models are generally in good agreement with the experimental results. Nevertheless, some deviations are seen between experimental and theoretical calculation results. The theoretical prediction models do not include any processing conditions such as moulding and curing. It is thought that these deviations have originated because of the missing processing parameters in theoretical prediction models. As a result of experimental studies, the lowest thermal conductivity value of expanded perlite-based panels was obtained as 43.5 mW/mK. Consequently, the heat transfer coefficient of the panels containing expanded perlite can be calculated nearly by the parallel method.

In order to assess the effectiveness of utilization of IOT and perlite in making bricks, a systematic research study was carried out.

6.2 EXPERIMENTAL INVESTIGATIONS

This section gives the details of laboratory investigations and pilot-scale studies, such as:

- Collection of required samples
- Determination of physico-chemical properties and sieve analysis in laboratory
- Casting of bricks with different composition

- Curing the moulded bricks for required duration
- Determination of various properties of finished bricks like unconfined compressive strength (UCS), durability, density, water absorption, thermal conductivity, etc.
- Pilot-scale study to assess the thermal efficiency of IOT–perlite bricks in comparison with conventional bricks

6.2.1 Collection of Samples

The materials used in the research work are IOT, perlite, sand and cement.

6.2.1.1 Iron Ore Tailings

IOT samples are collected from tailings dam of a Mining Company in Bellary district of Karnataka state using random sampling method (Figure 6.1) and safety precautions were taken by wearing protective gloves and mask. IOTs are the by-product produced in the beneficiation process of iron ore under certain technologies. Around 3 tonnes of IOT were collected and transported to the laboratory where the research work was being carried out. The mine management was kind enough to give the material free of cost.

6.2.1.2 Sand

Locally available river sand is used for present investigations as fine aggregates confirming that IS 2116: 1980 were used.

6.2.1.3 Cement

Ordinary Portland Cement (OPC) of Grade 53 (ACC) available in local market confirming to IS 12269:2013 is used in casting bricks. The cement is of uniform colour, i.e. grey with a light greenish shade and is free from any hard lumps.

FIGURE 6.1 Collecting iron ore tailings from a tailings pond.

6.2.1.4 Perlite

Perlite is procured from a local manufacturing company. The particles of perlite are chemically inert, very light weight, mildly abrasive and inorganic material with a porous structure. The quantity of perlite used for preparing bricks is 54 kg (i.e. nine bags containing 6 kg in each bag).

6.2.2 Physical and Chemical Properties of Samples

6.2.2.1 Iron Ore Tailings

Sieve analysis of IOT is carried out and the results are given in Table 6.1. A typical Particle-size distribution curve of the tailings is shown in Figure 6.2. The particles ranging from 4.75 mm and below are used as fine aggregates in making the tailings bricks. It is red in colour and has specific gravity of 2.71 and water absorption is 2.29%. Chemical compositions of tailings are presented in Table 6.2 and the elements present in IOT are given in Table 6.3.

6.2.2.2 Sand

Natural river sand is used for the studies. The water absorption and specific gravity of the sand are 0.1% and 2.65, respectively. Figure 6.3 shows the grain size distribution

TABLE 6.1
Particle Size Distribution of Iron Ore Tailings

Sieve Size	Weight (Wt.), %
Above 1.18 mm	21.40
600 μ to 1.18 mm	50.20
300 μ to 600 μ	13.20
15 μ to 300 μ	6.60
Below 15 μ	8.60
Average	100.00

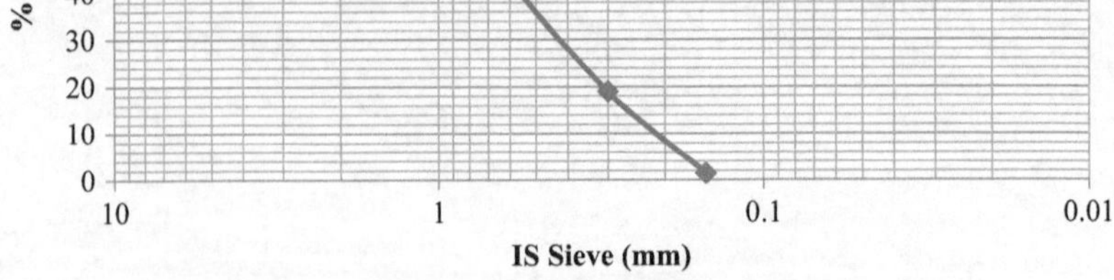

FIGURE 6.2 Particle size distribution of iron ore tailings.

TABLE 6.2
Chemical Composition of Iron Ore Tailings

Chemical Component in IOT	SiO_2	Al_2O_3	Fe_2O_3	CaO	MnO	K_2O	ZnO	CuO	PbO	LOI
% of component	56	10	8.3	4.3	1.7	1.5	0.1	0.2	0.4	3.3

TABLE 6.3
Percentage of Elements Present in Iron Ore Tailings

Element	Weight, %	Atomic Weight, %	Error, %
O	34.36	59.41	7.33
Al	9.01	9.24	8.58
Si	7.36	7.25	8.20
Pb	1.24	0.17	16.21
K	0.23	0.16	66.49
Ca	0.41	0.28	52.65
Mn	0.91	0.46	33.40
Fe	46.48	23.03	2.78

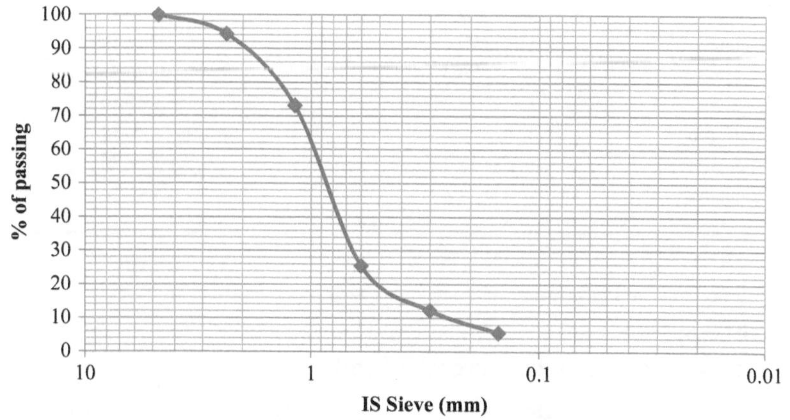

FIGURE 6.3 Grain size distribution of sand.

curve of sand. The maximum and minimum dry density is determined by using relative density equipment and is found to be 1870 kg/m^3 and 1500 kg/m^3, respectively.

The uniformity coefficient (Cu) is defined as a ratio and is calculated as the size opening that will just pass 60% of the sand (D_{60} value) divided by the size opening that will just pass 10% of the sand sample (D_{10} value). Coefficient of curvature is the parameter estimated using the gradation curve through sieve analysis. This parameter is used to classify the soil as well graded or poorly graded and is given by the relation as shown in Equation 6.1.

$$Cc = D^2_{30}/(D_{10} \times D_{60}) \tag{6.1}$$

The uniformity coefficient (Cu) and coefficient of curvature (Cc) found for the collected sand sample are 3.57 and 1.70, respectively. The values of Cu and Cc show that the sand is well graded.

The D_{50} is the size in microns that splits the distribution with half above and half below this diameter, or the portion of particles with diameters smaller and larger than this value are 50%, and this is also known as the median diameter. The median diameter (D_{50}) of sand used in manufacturing of brick is 0.89.

From sieve analysis, it is noticed that collected river sand percentage passing lies between the given percentage of passing in Zone I of sand grading chart. Therefore, Zone I sand passing through 4.75 mm sieve is used for casting of bricks (Table 6.4). The elements present in sand are given in Table 6.5.

TABLE 6.4
Sieve Analysis and Grading of Sand

IS Sieve	Percentage of Passing of Collected River Sand	Percentage Passing for Grading Zone I
10 mm	100	100
4.75 mm	100	90–100
2.36 mm	94.3	60–95
1.18 mm	73.2	30–70
600 micron	25.6	15–34
300 micron	12.2	5–20
150 micron	5.6	0–10

TABLE 6.5
Percentage of Elements Present in Sand

Element	Weight, %	Atomic Weight, %	Error, %
O	43.29	58.16	8.87
Na	0.67	0.62	35.14
Mg	0.88	0.78	16.13
Al	4.82	3.84	7.05
Si	44.46	34.03	4.01
K	0.66	0.36	32.43
Ca	1.26	0.68	19.91
Fe	3.96	1.53	11.61

6.2.2.3 Cement

Portland cement is used for research studies. OPC of grade53 conforming to IS 12269:1987 was used. It is grey in colour; particle size ranges from 0.007 to 0.2 mm and has a specific gravity of 3.15. The chemical composition and physical properties of cement are taken from manufacturer and the details are given in Table 6.6 and Table 6.7, respectively (www.acchelp.in).

TABLE 6.6
Chemical Composition of Ordinary Portland Cement

Sl. No.	Chemical Composition	Results	Requirements as per IS: 12269: 1987
1.	Soluble Silica, %	21.4	–
2.	Alumina, %	5.1	–
3.	Iron Oxide, %	3.6	–
4.	Lime, %	63.8	–
5.	Magnesia, %	0.8	Not more than 6.0%
6.	Insoluble Residue, %	0.8	Not more than 3.0%
7	Sulphur calculated as SO_3, %	2.3	Not more than 2.5 if C_3A is 5.0 or less & not more than 3.0 if C_3A is more than 5.0%
8.	Loss on Ignition, %	1.6	Not more than 4.0%
9.	Lime Saturation factor	0.91	Between 0.80 & 1.02
10	Proportion of Alumina to Iron Oxide	1.42	Not less than 0.66
11	Tri Calcium Aluminate	7.42	–
12	Chloride, %	0.022	Not more than 0.1%

TABLE 6.7
Physical Properties of Ordinary Portland Cement

Sl. No.	Physical Properties	Results	Requirements as per IS: 12269: 1987
1.	Fineness: Specific Surface, m^2/kg	349	Not less than 225
2.	Compressive Strength, MPa: at 3 Days 7 Days 28 Days	42.0 52.0 65.0	Not less than 27 Not less than 37 Not less than 53
3.	Setting Time, minutes Initial Final	195 280	Not less than 30 Not more than 600
4.	Soundness: Le-Chatelier Expn., mm Autoclave Expn., %	1.0 0.06	Not more than 10 mm Not more than 0.8%
5.	Normal Consistency, %	28.0	----------

TABLE 6.8
Typical Chemical Composition of Perlite

Chemical Composition		%
Silicon Dioxide	SiO_2	71–75
Alumina	Al_2O_3	12.5–18
Iron oxide	Fe_2O_3	0.5–1.5
Magnesium oxide	MgO	0.5–1.5
Quick lime	CaO	0.5–2.0
Caustic soda	Na_2O	2.9–4
Potassium oxide	K_2O	4–5
Hardness on Mohr's scale		3–4

TABLE 6.9
Typical Physical Properties of Perlite

S. No	Property	Value
1	Colour	Whitish grey
2	Apparent density, kg/m^3	150
3	Specific Gravity	2.3
4	pH	6.5
5	Water absorption, % of mass	250
6	Thermal conductivity, W/mK	0.043

6.2.2.4 Perlite

Perlite is procured from Keltech Energies Ltd and its properties are provided by the manufacturer (www.keltechenergies.com). The chemical composition of perlite is given in Table 6.8. Physical properties of Perlite are given in Table 6.9. The elements present in perlite are given in Table 6.10.

6.2.3 Mix Proportion

The materials for manufacturing the brick consist of IOT, sand, cement and perlite. To make the IOT–perlite brick following mix proportion are arrived by trial and error method. Table 6.11 shows the various mix proportions. The mixture contains perlite as additives, whereas added in percentage of 0, 2 and 5. Initially, casting of bricks is considered without any perlite. Cement percentage is fixed at 10%, and from remaining 90%, sand is replaced with IOT, 30 to 60%, with 10% incremental order. Similarly, in the next stage, cement percentage is increased to 15% and later 20%, and the same process is followed. Later perlite is increased to 2% and 5% and the same process is followed as shown in the Table 6.11.

TABLE 6.10
Percentage of Elements Present in Perlite

Element	Weight, %	Atomic Weight, %	Error, %
O	48.39	62.56	8.65
Na	3.76	3.38	11.61
Mg	0.79	0.67	15.60
Al	7.56	5.79	5.95
Si	33.19	24.44	4.44
K	4.22	2.23	7.70
Ca	0.78	0.40	25.80
Ti	0.55	0.24	34.00
Fe	0.76	0.28	38.29

Volume of one brick is 0.00199 m^3. The weight of each raw material is calculated by multiplying its percentage in the brick to its density which is given in Table 6.12. The quantity of each raw material is calculated as:

$$\text{Quantity of material} = \text{Volume of the brick} \times \text{percentage of the raw material} \times \text{density} \quad (6.2)$$

6.2.4 Preparation of Bricks

Thorough mixing of the materials is essential for the production of uniform product. The mixing should ensure that the mass becomes homogeneous, uniform in colour and consistency. Normally, a batch mix made with ingredients corresponds to different percentage of IOTs, sand, perlite and cement.

The normal hand mould is used to cast the bricks in the laboratory with the standard size of 23.0 cm length, 11.25 cm width and 7.5 cm height (Figure 6.4). They are casted according to the standard procedure with various mix proportions arrived. The required quantity of IOTs, Sand, Cement and Perlite is calculated previously; accordingly, the materials are mixed.

Then the required quantity of water is added and mixed thoroughly. The prepared mix is poured into the mould, and it is compacted. After casting, all the test specimens are kept at room temperature for 24 hours and then demoulded. These are then placed in water curing tank.

Curing is one of the most essential parts of brick production. Optimum curing duration has a major effect on the quality of the end product. The curing of brick is most important as it is very essential for keeping the hydration process of cement until brick attains the maximum compressive strength, which increases but slowly after 21 days from initial placing time. Hence, the bricks are cured for 21 days in water tank. Bricks are casted in six numbers for each varying proportion of IOT, cement, sand, perlite and water combination (Table 6.13).

TABLE 6.11
Various Mix Proportions for Bricks

Proportion	Iron Ore Tailings, %	Sand, %	Cement, %	Perlite, %
1	30	60	10	0
2	40	50		
3	50	40		
4	60	30		
5	30	58	10	2
6	40	48		
7	50	38	10	2
8	60	28		
9	30	55	10	5
10	40	45		
11	50	35		
12	60	25		
13	30	55	15	0
14	40	45		
15	50	35		
16	60	25		
17	30	53	15	2
18	40	43		
19	50	33		
20	60	23		
21	30	50	15	5
22	40	40		
23	50	30		
24	60	20		
25	30	50	20	0
26	40	40		
27	50	30		
28	60	20		
29	30	48	20	2
30	40	38		
31	50	28		
32	60	18		
33	30	45	20	5
34	40	35		
35	50	25		
36	60	15		
37	50	23	20	7

TABLE 6.12
Quantity of Materials Used in Each Brick

Proportions	Iron Ore Tailings, kg	Sand, kg	Cement, kg	Perlite, kg
1	1.07	2.089	0.318	0
2	1.43	1.741	0.318	0
3	1.79	1.393	0.318	0
4	2.15	1.045	0.318	0
5	1.07	2.020	0.318	0.006
6	1.43	1.672	0.318	0.006
7	1.79	1.323	0.318	0.006
8	2.15	0.975	0.318	0.006
9	1.07	1.915	0.318	0.015
10	1.43	1.567	0.318	0.015
11	1.79	1.219	0.318	0.015
12	2.15	0.871	0.318	0.015
13	1.07	1.915	0.477	0
14	1.43	1.567	0.477	0
15	1.79	1.219	0.477	0
16	2.15	0.871	0.477	0
17	1.07	1.846	0.477	0.006
18	1.43	1.497	0.477	0.006
19	1.79	1.149	0.477	0.006
20	2.15	0.801	0.477	0.006
21	1.07	1.741	0.477	0.015
22	1.43	1.393	0.477	0.015
23	1.79	1.045	0.477	0.015
24	2.15	0.696	0.477	0.015
25	1.07	1.741	0.637	0
26	1.43	1.393	0.637	0
27	1.79	1.045	0.637	0
28	2.15	0.696	0.637	0
29	1.07	1.672	0.637	0.006
30	1.43	1.323	0.637	0.006
31	1.79	0.975	0.637	0.006
32	2.15	0.627	0.637	0.006
33	1.07	1.567	0.637	0.015
34	1.43	1.219	0.637	0.015
35	1.79	0.871	0.637	0.015
36	2.15	0.522	0.637	0.015
37	1.79	0.801	0.637	0.021

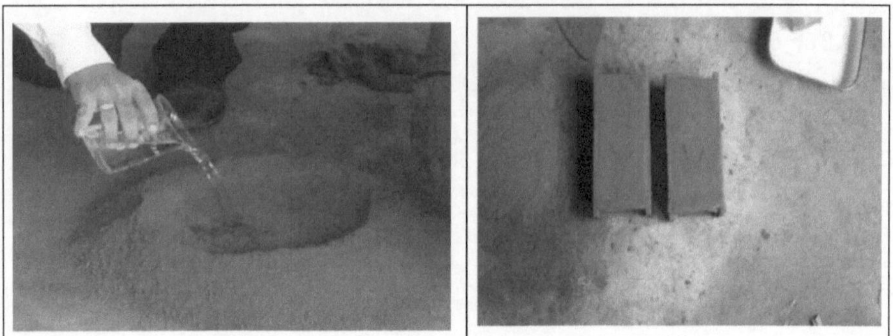

Mixing Moulding

FIGURE 6.4 A view of casting of bricks.

Accordingly for 10 and 15% of cement, 336 samples are casted, whereas for 20% cement, 364 samples are casted. In total, 1036 samples are casted. In fact, more than 1500 samples are casted and some of them are wasted during different tests. For pilot-scale study, further 1000 bricks are casted.

Bricks with 0%, 2% and 5% of perlite and 15% of cement as constant for various proportions of IOT and sand are casted. These bricks are tested for water absorption, compressive strength and thermal conductivity. Details of number of samples casted in each case for 15% of cement are given in Table 6.14.

Bricks with 20% cement content and with 0%, 2% and 5% perlite were casted (Table 6.15). Water absorption, compressive strength and thermal conductivity were tested.

6.3 LABORATORY INVESTIGATIONS

Quality assessment and durability tests are carried out to obtain optimum percentage of raw materials. After curing, the bricks are subjected to quality assessment tests like UCS, durability tests such as water absorption and thermal conductivity. Density of brick is determined to check its weight and degree of compactness of material and efflorescence test is done to ascertain its suitability to retain the aesthetic appearance of the building.

Brick production is designed for different proportions. IOT were varied for percentage of 30, 40, 50 and 60 by keeping cement as constant 10%, 15% and 20% and perlite as 0%, 2% and 5% for all proportions of IOT.

6.3.1 Density

Density is also called unit weight of substance. Density represents the degree of compactness of material. If the material is of higher density, it is more compacted material. Density of construction materials are its mass per unit volume. Its value of construction material will also help to find out the quantity of material needed

TABLE 6.13
Samples Casted For various Experiments by Retaining 10% Cement as Constant

		Number of Samples Casted with 10% Cement Content											
Name of the test	**Curing period in days**	**A-30% B-60% C-0%**	**A-40% B-50% C-0%**	**A-50% B-40% C-0%**	**A-60% B-30% C-0%**	**A-30% B-58% C-2%**	**A-40% B-48% C-2%**	**A-50% B-38% C-2%**	**A-60% B-28% C-2%**	**A-30% B-55% C-5%**	**A-40% B-45% C-5%**	**A-50% B-35% C-5%**	**A-60% B-25% C-5%**
Water absorption	21	6	6	6	6	6	6	6	6	6	6	6	6
Compressive strength	21	6	6	6	6	6	6	6	6	6	6	6	6
Thermal conductivity	21	16	16	16	16	16	16	16	16	16	16	16	16
Sub Total		28	28	28	28	28	28	28	28	28	28	28	28
Total								336					

Notation: A – IOT, B – sand, C – Perlite

TABLE 6.14
Samples Casted for Various Experiments by Retaining 15% Cement as Constant

Name of the test	Curing period in days	Number of samples casted with 15% cement content											
		A-30% B-55% C-0%	A-40% B-45% C-0%	A-50% B-35% C-0%	A-60% B-25% C-0%	A-30% B-53% C-2%	A-40% B-43% C-2%	A-50% B-33% C-2%	A-60% B-23% C-2%	A-30% B-50% C-5%	A-40% B-40% C-5%	A-50% B-30% C-5%	A-60% B-20% C-5%
Water absorption	21	6	6	6	6	6	6	6	6	6	6	6	6
Compressive strength	21	6	6	6	6	6	6	6	6	6	6	6	6
Thermal conductivity	21	16	16	16	16	16	16	16	16	16	16	16	16
Sub Total		28	28	28	28	28	28	28	28	28	28	28	28
Total		336											

Notation: A – IOT, B – sand, C – Perlite

TABLE 6.15
Samples Casted for Various Experiments by Retaining 20% Cement as Constant

Name of the test	Curing period in days	Number of samples casted with 20% cement content												
		A-30% B-50% C-0%	A-40% B-40% C-0%	A-50% B-30% C-0%	A-60% B-20% C-0%	A-30% B-48% C-2%	A-40% B-38% C-2%	A-50% B-28% C-2%	A-60% B-18% C-2%	A-30% B-45% C-5%	A-40% B-35% C-5%	A-50% B-25% C-5%	A-60% B-15% C-5%	A-50% B-23% C-7%
Water absorption	21	6	6	6	6	6	6	6	6	6	6	6	6	6
Compressive strength	21	6	6	6	6	6	6	6	6	6	6	6	6	6
Thermal conductivity	21	16	16	16	16	16	16	16	16	16	16	16	16	16
Sub Total		28	28	28	28	28	28	28	28	28	28	28	28	28
Total								364						
Grand Total								336 + 336+ 364 = 1036						

Notation: A – IOT, B – sand, C – Perlite

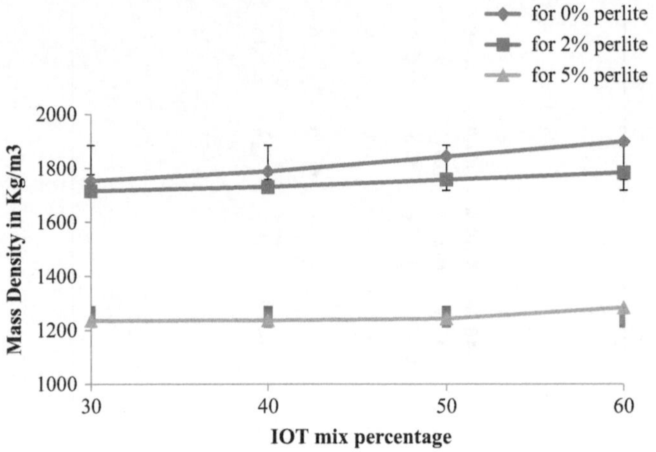

FIGURE 6.5 Density of bricks with 10% cement for different IOTs mix percentage.

for particular space. It also helps to know the dead load acting on the building. It is expressed in kg/m^3 and shows compactness of building material.

Weight and density of bricks of size 230 mm × 112.5 mm × 75 mm are determined as per IS 1077:1963 and the average of at least six samples is reported. The brick placed on the dry cloth and gently surface dried with the cloth and transferred it to the second dry cloth, if the first cannot remove moisture further. Then it is exposed to the atmosphere away from direct sunlight or any other source of heat for not less than 10 minutes or until it appears to be completely surface dry and then weight of individual brick is recorded as per IS 1077:1963. On the basis of weight and volume, density of bricks is calculated and only average dry density is taken for analysis.

Figure 6.5 shows when IOT in bricks increased from 30% to 60%, density also increased gradually (1754.32 kg/m^3–1898.87 kg/m^3), later it is slightly/gradually reduced for all percentages of perlite at a fixed percentage of 10% cement content, i.e. from 0% to 5% of perlite, density reduced from 1843.29 kg/m^3 to 1242.89 kg/m^3. Similar trend is observed in Figures 6.6 and 6.7 for 15% and 20% cement content, respectively. So, it can be concluded that addition of perlite drastically decreases the density of brick.

6.3.2 Water Absorption

The bricks are weighed after immersing in water for 24 hours and later placed in oven till it reaches a stage of constant weight and based on initial and final weight, water content is determined. The water absorption test is performed according to IS 3495(part II): 1992. The percentage of water absorption (w_a) is calculated using the following formula:

$$w_a = \frac{w_2 - w_1}{w_1} \times 100 \tag{6.3}$$

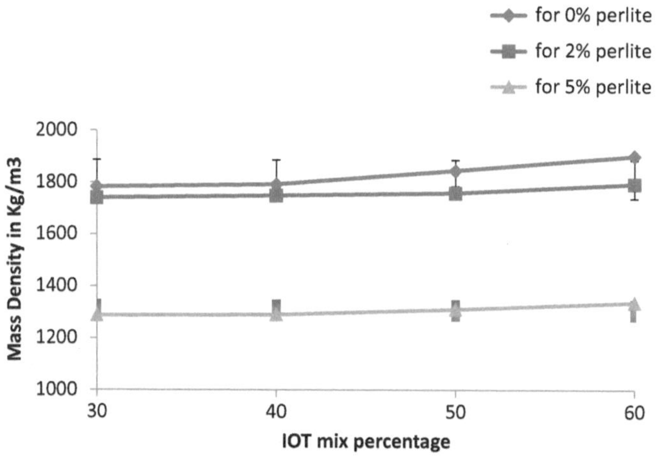

FIGURE 6.6 Density of bricks with 15% cement for different IOTs mix percentage.

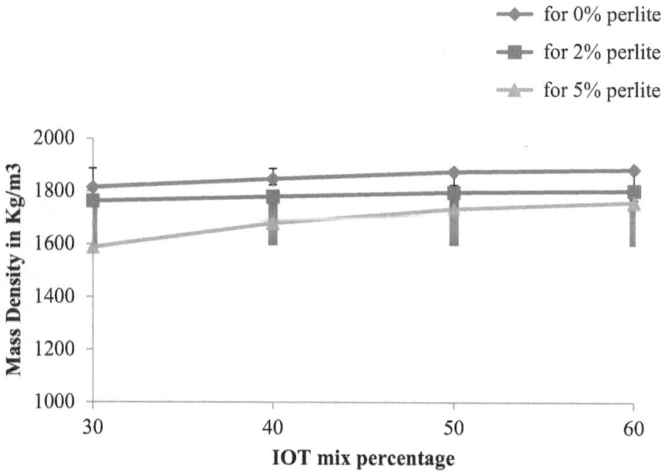

FIGURE 6.7 Density of bricks with 20% cement for different IOTs mix percentage.

where, w_1 is the weight of the dry brick after keeping in oven for 24 hours at 100°C temperature; and w_2 is the weight of the brick after curing at the specified temperature.

Water absorption tests on bricks are conducted to determine durability property of bricks such as degree of burning, quality and behaviour of bricks in weathering. A brick with water absorption of less than 7% provides better resistance to damage by freezing. The degree of compactness of bricks can be obtained by water absorption test, as water is absorbed by pores in bricks. Water absorption test is conducted as per IS 3495:1992 Part II, which gives the percentage of water absorbed by bricks in

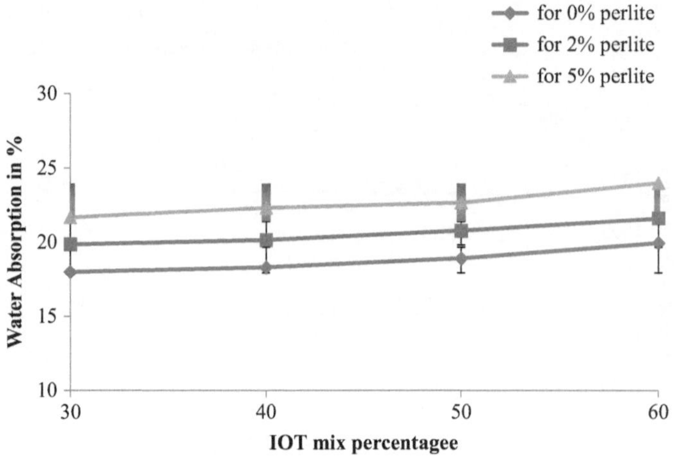

FIGURE 6.8 Water absorption of bricks for 10% cement for different IOTs mix percentage.

24 hours. Dry weights of six brick specimens of different percentage of IOT, sand, perlite and cement brick specimens are taken. All the specimens are immersed in water at the same time. The wet weight of specimens is recorded and the average percentage of water absorption is calculated for each different composition of specimen.

According to IS3495 (Part II): 1992, water absorption should not be more than 20% for 24 hours immersion. Slight increase in water absorption is noticed when there is an increase in IOT and higher percentage of water absorption is found when perlite is added as it is a porous structure. When cement addition is increased from 10% to 20% in the interval of 5%, water absorption decreased drastically. Figure 6.8 shows the trend of water absorption for 10% of cement content, with increase in IOT percentage. Figures 6.9 and 6.10 also show, as IOT increased from 30 to 60%, water absorption also increased due to the increase in surface area of IOT. Although the perlite content goes on increasing (0%–5%), water absorption gradually increased as perlite has a relatively high water content and at 20% cement addition water absorption was favourable by satisfying the IS code.

6.3.3 Compressive Strength

The Uni-axial Compressive Strength (UCS) is tested using a compression testing machine as per IS 3495 (part I):1992 (Figure 6.11). Six bricks are tested in each case for the UCS and the average of the UCS values is taken for analysis purpose. Ultimate load at which the specimens failed is noted down to find the compressive strength. Rupture surface of the specimens after failure are visually inspected. Figure 6.11 shows a view of a specimen after failure in compression. The uni-axial compressive strength of the brick is calculated using the following formula:

$$\text{UCS} = \frac{\text{Load}\left(P\right)}{\text{Cross-sectional area}\left(A\right)\text{of the brick}} \tag{6.4}$$

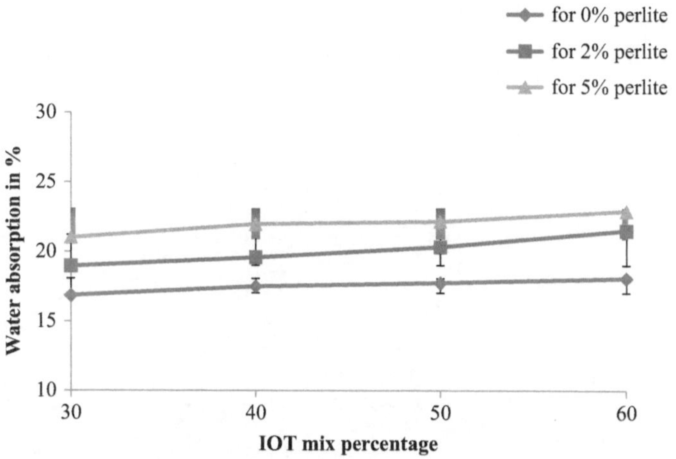

FIGURE 6.9 Water absorption of bricks for 15% cement for different IOTs mix percentage.

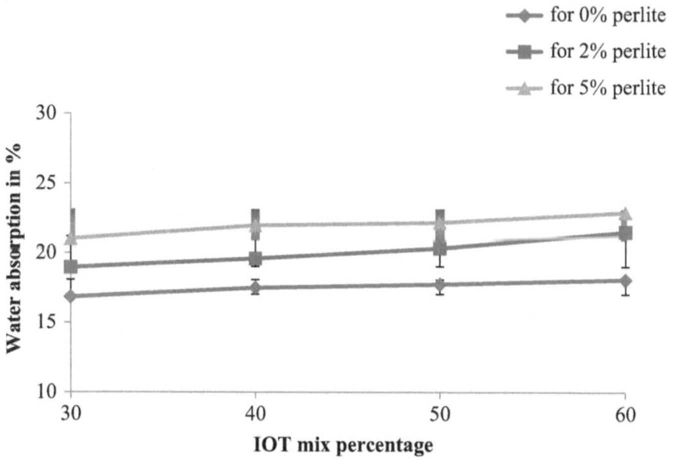

FIGURE 6.10 Water absorption of bricks for 20% cement for different IOTs mix percentage.

Compressive strength test on bricks is carried out to determine the load carrying capacity of bricks under compression with the help of compression testing machine. Bricks are generally used for the construction of load-bearing masonry walls, columns and footings. These load-bearing masonry structures experience mostly the compressive loads. Thus, it is important to know the compressive strength of bricks to check for its suitability for construction. Cubic specimens are tested in compression-testing machine after specified curing period in accordance with the IS specifications.

Compressive strength is the more common performance parameter measured and used to study the properties of brick. It gives an idea about the mechanical properties

FIGURE 6.11 Illustration of compression test on IOT–perlite brick.

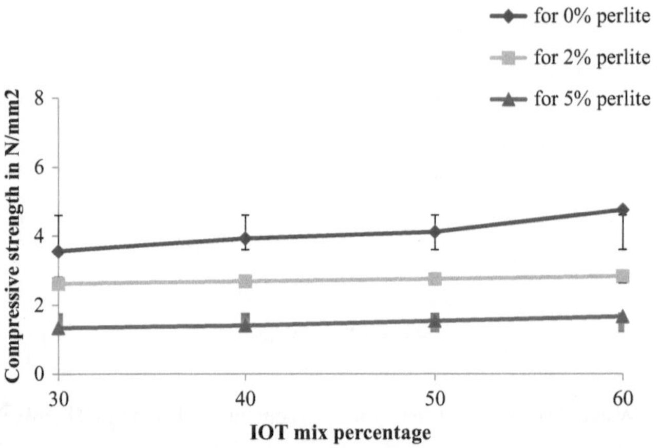

FIGURE 6.12 Compressive strength of bricks with 10% cement for different IOTs mix percentage.

of brick. As per IS 3495:1992 Part I, the minimum compressive strength of bricks should be 3.5 MPa. Figure 6.12 shows that the compressive strength increased as the IOTs percentage increased from 30% to 60%, at a fixed percentage of 10% cement content. But as the perlite percentage increased from 0% to 5%, compressive strength decreased for all IOTs mixes. Similar trend is observed in Figures 6.13 and 6.14 for 15% and 20% cement content, respectively. So, it can be stated that maximum of

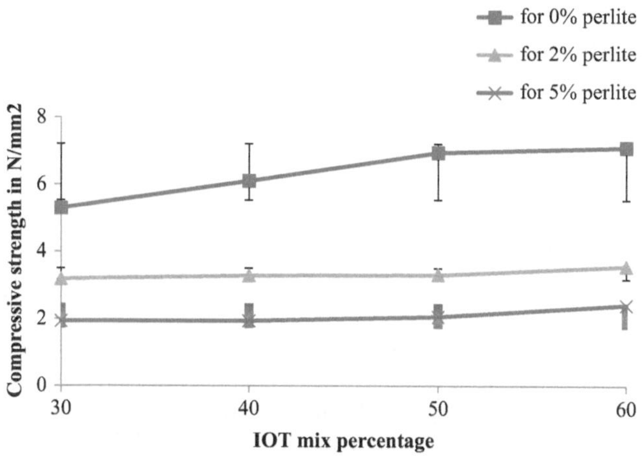

FIGURE 6.13 Compressive strength of bricks with 15% cement for different IOTs mix percentage.

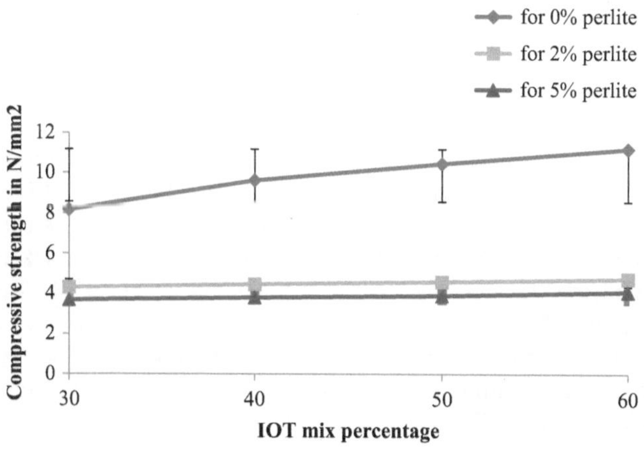

FIGURE 6.14 Compressive strength of bricks with 20% cement for different IOTs mix percentage.

50% IOTs mix with 20% cement gives better compressive strength and 5% perlite addition will not compromise on the required compressive strength. Though the perlite addition reduces compressive strength but reduces density and thermal conductivity, which are advantageous.

In fact, a few bricks were casted with 20% cement combination and 7% perlite also, but they have resulted less than IS code required compressive strength, i.e. 3.5 MPa. So, the influence of 7% perlite is not included in the study.

6.3.4 Efflorescence

Efflorescence is a crystalline, salty deposit that occurs on the surfaces of bricks. It is white, sometimes a brilliant white or an off white colour. Efflorescence has been a real bug bear of the building and construction industry for many years. The formation of these salt deposits is not an inexplicable phenomenon; they are simply water-soluble salts that come from different sources to ruin the looks of the building. The presence of efflorescence causes disagreeable appearance, damp patches on wall, and will even detach small fragments of the materials composing the wall. The liability of efflorescence shall be reported as nil, slight, moderate, heavy or serious in accordance as per IS 3495 (Part 3): 1992.

Efflorescence test is carried out as per IS 3495:1992 Part III. The result obtained for this test is 'Nil', when brick is containing 5% of perlite, 50% of IOTs and 20% cement, almost no perceptible deposit of efflorescence is observed. When there is no perlite in brick, the presence of efflorescence is observed. While efflorescence is not a risk structurally, it affects the appearance of a building; particularly, it features coloured brickwork. It concludes that presence of perlite in brick shows no efflorescence, which can be used for brickwork and reduces the maintenance cost, which is an added advantage.

6.3.5 Thermal Conductivity

The property that characterizes the ability of a material to transfer heat is thermal conductivity (k). It is a specific property of the material (k), which is a measure of the rate at which heat (energy) passes perpendicularly through a unit area of a homogenous material of unit thickness for a temperature difference of 1°C. Thermal conductivity measurement is important to understand the heat flow in bricks. There are two main methods to measure thermal conductivity of materials, viz. the steady-state method and the transient method (Bindiganavile et al. 2012). Steady-state methods are adopted for homogenous materials. In this method, the flux is proportional to the temperature gradient along the direction of flow. The experimental procedures are time consuming. However, the thermal conductivity values obtained by this method are accurate. The method of steady-state thermal conductivity analysis includes guarded hot plate method. The thermal conductivity test is performed according to the American Society for Testing & Materials (ASTM) C-177 and IS-3346.

The essential parts of Guarded Hot Plates are the following: the hot plates, the cold plates, the heater assembly, thermocouples and the specimens in position.

One-dimensional heat flow through a flat specimen, an arrangement for maintaining its faces at constant temperature and a measure of heat flow through a known area for the measurement of thermal conductivity (k) is required. To eliminate the distortion caused by edge losses in unidirectional heat flow, the central plate is surrounded by a guard ring, which is separately heated. Temperatures are measured by calibrated thermocouples either attached to the plates or to the specimen at the hot and cold

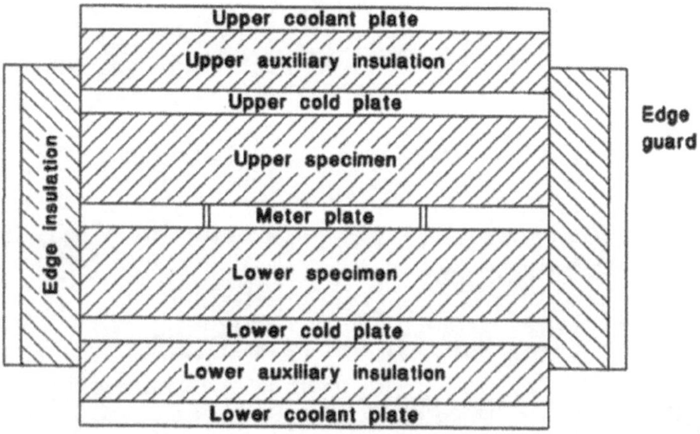

FIGURE 6.15 A typical high-temperature guarded hot plate apparatus (cross-section view).

faces. Figure 6.15 shows the overall layout of a typical high-temperature guarded hot plate apparatus.

Knowing the heat input to the central plate heater, the temperature difference across the specimen, its thickness and the area of the specimen, k, can be calculated by using the following formula:

$$k = \frac{Q}{2A} \times \frac{L}{T_h - T_c} \text{ Watt/m°C} \tag{6.5}$$

where,

k = Thermal Conductivity of sample, W/m°C or W/mk
Q = Heat flow rate in the specimen, Watts
A = Metering area of the specimen, m^2
T_h = Hot plate temperature, °C
T_c = Cold plate temperature, °C
L = Thickness of specimen, m

If the specimen thicknesses are different and the respective hot and cold temperatures are different, then,

$$k = Q \times \left\{ \frac{\frac{1}{T_{h1} - T_{c1}}}{T_{h1} - T_{c1}} \quad \frac{\frac{1}{T_{h2} - T_{c2}}}{L_2} \right\} \text{ W/mK (In S.I.units)} \tag{6.6}$$

Thermal conductivity test is conducted by placing the specimen on either side of the heating plate assembly, uniformly touching the cooling plates (Figure 6.16).

FIGURE 6.16 A view of thermal conductivity apparatus.

Then the outer container is filled with loose fill insulation such as glass wool (supplied in small cloth packets). The cooling circuit is started, known input is given to central and guard heaters through separate single-phase supply lines with a dimmer stat in each line and it is adjusted to maintain the desired temperature. The guard heater input is adjusted in such a way that there is no radial heat flow, which is checked from thermocouple readings and is adjusted accordingly. Power input to the central heater and the thermocouple readings are recorded after every 10 minutes till a reasonably steady-state condition is reached. The final steady-state values are taken for calculations.

A pilot-scale study is important to ensure the quality and efficiency of the laboratory study. This study focuses on the feasibility study on IOT and perlite in bricks. It also determines physical and chemical properties of IOT and perlite and how its reaction process affects the bricks.

Raw materials are tested in laboratory to ensure that the raw materials used in brick products are suitable for their intended use. Conducting analysis on raw materials using appropriate test methods and successfully meeting the challenges of such testing can prevent costly production problems and delays by confirming early in the production process. The laboratory testing of raw materials used in IOT–perlite brick appeared to tend towards the higher properties than the ordinary bricks. So, a pilot-scale study is taken up based on positive results of laboratory studies.

Thermal conductivity is the material property that determines how fast or how much heat can be transferred through a material for a given temperature difference. Accurate information of a material's thermal conductivity is crucial in heat transfer applications. The thermal conductivity is not always constant. The main factors affecting the thermal conductivity are the density of material, moisture of material and ambient temperature. With increasing density, moisture and temperature, the thermal conductivity also increases.

The dense solid materials (IOT) tend to have high levels of conductivity, whereas materials with very small matter and large proportion of voids (gas or air bubbles, not large enough to carry heat by convection, i.e. perlite) have the lowest conductivities (www.researchgate.net). Thermal conductivity test is carried out as per ASTM C-177 and IS-3346 on bricks with different combination of materials. The variation of thermal conductivity of samples with IOT, sand, perlite and cement is given in Table 6.16.

Thermal conductivity tests are carried out simultaneously with other tests. Many trials are done in order to confirm the influence of perlite. In general, it is found that as perlite percentage increases, the thermal conductivity decreases. Among the tests conducted, the highest value of the thermal conductivity, i.e. 1.7847 W/mk, is obtained for samples produced with 60% IOT, 20% sand, 0% perlite and 20% cement. As perlite content increased, thermal conductivity decreased. For 0%, 2% and 5% perlite, the reductions in thermal conductivity are noted as 1.778 W/mk, 1.285 W/mk, and 0.905 W/mk, respectively, at 30% IOT and 20% cement. In the entire mix, thermal conductivity reduced, when perlite percentage increased from 0 to 5% for all IOT percentages and cement. This describes that the thermal conductivity will decrease when there is increase in perlite content.

Figure 6.17 shows that the thermal conductivity increased as the IOT percentage increased from 30% to 60%; later it is reduced for all percentages (0%–5%) of perlite at a fixed percentage of 10% cement content. As the perlite percentage increased from 0% to 5%, thermal conductivity decreased for all IOT mixes. Similar trend is observed in Figures 6.18 and 6.19 for 15% and 20% cement content, respectively. So, it can be stated that perlite is the best thermal insulator which have the lowest thermal conductivity and combination of 50% IOT mix, 5% perlite and 20% cement addition will give better thermal conductivity by keeping in view of other properties such as density, compressive strength and water absorption.

Based on laboratory-scale studies, it can be concluded that 50:25:20:5 (IOT: sand: cement: perlite) mix gives optimum results. So, in order to prove the effectiveness of optimum mix bricks, a pilot-scale study is carried out.

These observations are in line with the findings of Arunraja et al. (2017) with lightweight bricks of size 190 mm × 90 mm × 90 mm using perlite and lime in various ratios like 70:30, 75:25, 80:20 and 85:15. This has resulted 3.3 MPa compressive strength, and the present study obtained 3.89 MPa which satisfies the IS code.

TABLE 6.16
Thermal Conductivity Values for Different Combination of Additives

Sl. No	Percentage of Raw materials, % IOTs	Sand	Perlite	Cement	Avg. Thermal conductivity, W/mk	Decrease in percentage of Thermal Conductivity with reference to IOT-60%, Sand-20%,Perlite-0% and cement-20%
1.	30	60	0	10	1.491	16.47
2.	40	50			1.495	16.25
3.	50	40			1.497	16.13
4.	60	30			1.499	16.02
5.	30	58	2	10	1.220	31.65
6.	40	48			1.236	30.76
7.	50	38			1.259	29.46
8.	60	28			1.290	27.73
9.	30	55	5	10	0.901	49.52
10.	40	45			0.903	49.41
11.	50	35			0.912	48.91
12.	60	25			0.924	48.23
13.	30	55	0	15	1.492	16.41
14.	40	45			1.497	16.13
15.	50	35			1.514	15.18
16.	60	25			1.521	14.79
17.	30	53	2	15	1.271	28.79
18.	40	43			1.278	28.40
19.	50	33			1.291	27.67
20.	60	23			1.310	26.61
21.	30	50	5	15	0.903	49.41
22.	40	40			0.904	49.35
23.	50	30			0.939	47.74
24.	60	20			0.939	47.74
25.	30	50	0	20	1.778	0.39
26.	40	40			1.780	0.28
27.	50	30			1.782	0.17
28.	60	20			1.785	0
29.	30	48	2	20	1.285	28.01
30.	40	38			1.289	27.78
31.	50	28			1.292	27.62
32.	60	18			1.312	26.49
33.	30	45	5	20	0.905	49.30
34.	40	35			0.906	49.24
35.	50	25			0.920	48.46
36.	60	15			0.941	47.28
37.	50	23	7	20	0.7843	56.06

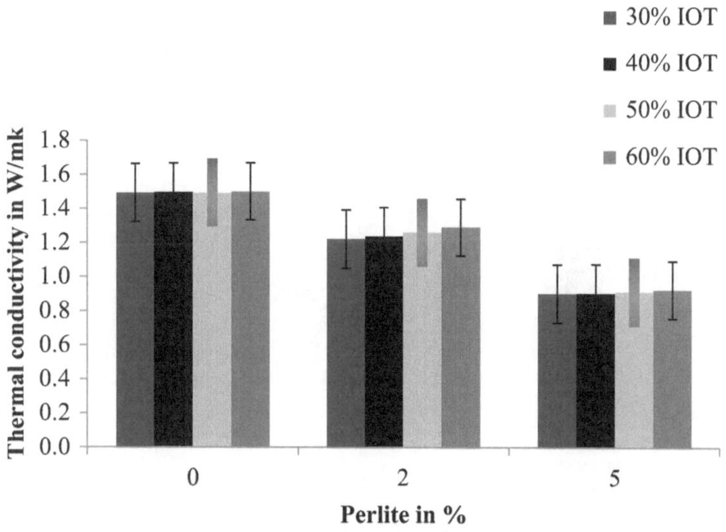

FIGURE 6.17 Thermal conductivity of bricks with 10% cement for different IOTs mix percentage.

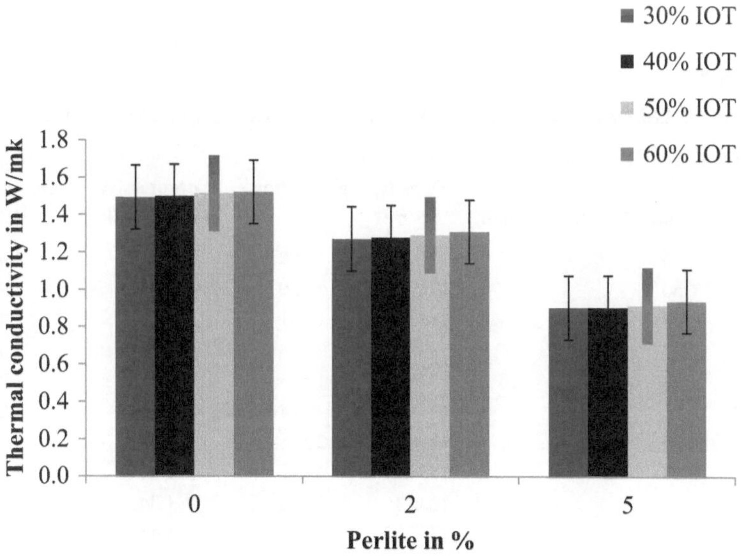

FIGURE 6.18 Thermal conductivity of bricks with 15% cement for different IOTs mix percentage.

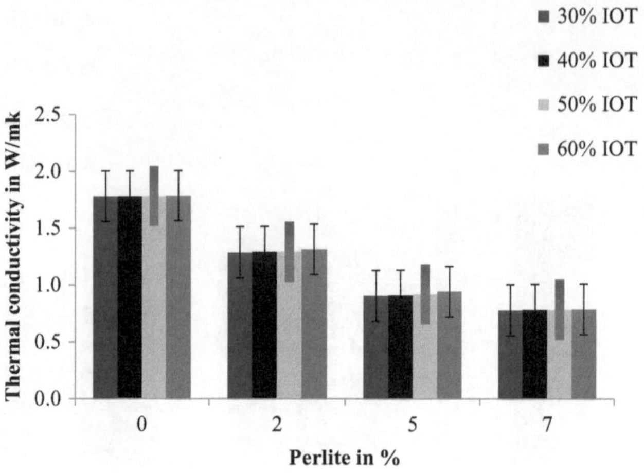

FIGURE 6.19 Thermal conductivity of bricks with 20% cement for different IOTs mix percentage.

6.3.6 Summary of All Properties

The summary of all tests carried out in laboratory are given in Table 6.17.

TABLE 6.17
Outcome of Laboratory Studies Conducted for the IOT–Perlite Brick

Sl. No	Percentage of Raw materials, % IOTs	Sand	Perlite	Cement	Density, kg/m³	Water absorption, %	Compressive strength, MPa	Thermal conductivity, W/mk
1.	30	60	0	10	1754.32	17.99	3.56	1.491
2.	40	50			1789.11	18.30	3.93	1.495
3.	50	40			1843.29	18.91	4.12	1.497
4.	60	30			1898.87	21.30	4.76	1.499
5.	30	58	2	10	1716.45	19.85	2.62	1.220
6.	40	48			1731.60	20.14	2.69	1.236
7.	50	38			1758.22	20.76	2.75	1.259
8.	60	28			1783.96	22.92	2.83	1.290
9.	30	55	5	10	1236.19	21.66	1.34	0.901
10.	40	45			1237.23	22.32	1.41	0.903
11.	50	35			1242.89	22.67	1.54	0.912
12.	60	25			1284.12	24.01	1.67	0.924
13.	30	55	0	15	1783.44	16.85	5.29	1.492
14.	40	45			1792.72	17.50	6.10	1.497
15.	50	35			1845.79	17.75	6.95	1.514
16.	60	25			1901.96	18.08	7.10	1.521

(Continued)

TABLE 6.17 (CONTINUED)
Outcome of Laboratory Studies Conducted for the IOT–Perlite Brick

17.	30	53	2	15	1740.67	18.96	3.18	1.271
18.	40	43			1749.45	19.58	3.28	1.278
19.	50	33			1759.32	20.32	3.30	1.291
20.	60	23			1794.27	22.67	3.56	1.310
21.	30	50	5	15	1287.73	21.00	1.93	0.903
22.	40	40			1290.30	21.96	1.93	0.904
23.	50	30			1310.92	22.16	2.06	0.916
24.	60	20			1337.19	22.90	2.41	0.939
25.	30	50	0	20	1813.85	8.09	8.16	1.778
26.	40	40			1845.79	8.29	9.62	1.780
27.	50	30			1873.12	9.12	10.45	1.782
28.	60	20			1902.48	9.56	11.19	1.784
29.	30	48	2	20	1762.31	13.21	4.30	1.285
30.	40	38			1778.81	13.53	4.45	1.289
31.	50	28			1795.81	14.06	4.58	1.292
32.	60	18			1853.53	14.61	4.72	1.312
33.	30	45	5	20	1587.36	14.25	3.68	0.905
34.	40	35			1678.32	14.50	3.79	0.906
35.	50	25			1723.43	14.82	3.89	0.920
36.	60	15			1758.19	15.64	4.08	0.941
37.	50	23	7	20	1103.76	15.60	1.79	0.7843

6.4 STUDY OF EFFICIENCY OF IOT–PERLITE BRICKS: A PILOT-SCALE STUDY

Pilot-scale study is carried out to analyse masonry building to assess the thermal efficiency and the variation of temperature induced in a structure under different intervals of time. Two rooms are constructed, one with IOT and perlite-mixed bricks and other with locally available conventional bricks, the description of the building analysed, the results obtained are presented in the following sections.

6.4.1 Description of the Model Rooms

To compare the results and to know the efficiency of the IOT–perlite bricks, two model rooms are constructed one of IOT–perlite bricks and other conventional fired bricks as shown in Figures 6.20 and 6.21, respectively. These two model rooms are built in similar conditions like in elevation, dimension and height. It is constructed nearby, exposing to the same intensity of light and free from all obstructions.

The room of conventional fired bricks constructed and pointed with cement mortar 1:4 and IOT–perlite brick room with mortar of IOT, sand, cement and perlite mix as of optimum ratio 50:25:20:5. Both the rooms are laid with plain cement concrete for floor and the roof is covered with asbestos cement sheet. The thickness of the wall is one brick thick having 60 cm wide door opening in east and the height of the room is 1.9 m with necessary slope.

FIGURE 6.20 A view of model room constructed with IOT and perlite-mixed bricks.

FIGURE 6.21 A view of model room constructed with ordinary bricks.

6.4.2 Experimental Procedure

The temperature is measured with two devices, i.e. Infrared Thermometer (model CENTER 350 Series) (www.primusthai.com) and Probe Temperature Meter (model V&A Instrument) (www.hertige.com). The wall temperature inside and outside of IOT–perlite brick wall and ordinary brick walls are measured at morning, afternoon and evening by Infrared Thermometer (Figure 6.22). Atmospheric temperature and the temperature inside the rooms are measured at different intervals of time using Probe Temperature Meter (Figure 6.23). Both of these temperature metres are calibrated with reference to the temperature of human body and water (Figure 6.24).

To measure temperature using infrared thermometer, the unit is pointed at an object and the trigger is pulled. Distance to spot size ratio and field of view is considered. The laser is used for aiming the target for reference. The temperature reading is updated on the LCD. When the trigger is released, the reading will automatically display on the LCD for 10 more seconds. After 10 seconds, this thermometer will power down itself to save its battery. The temperature which is displayed in LCD is

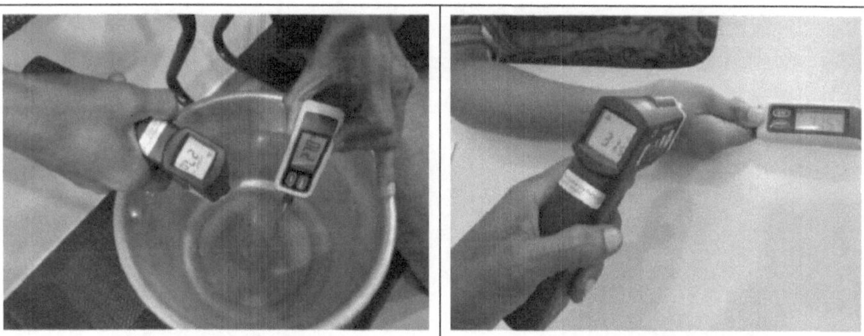

Temperature of water	Temperature of human body

FIGURE 6.22 Infrared thermometer.

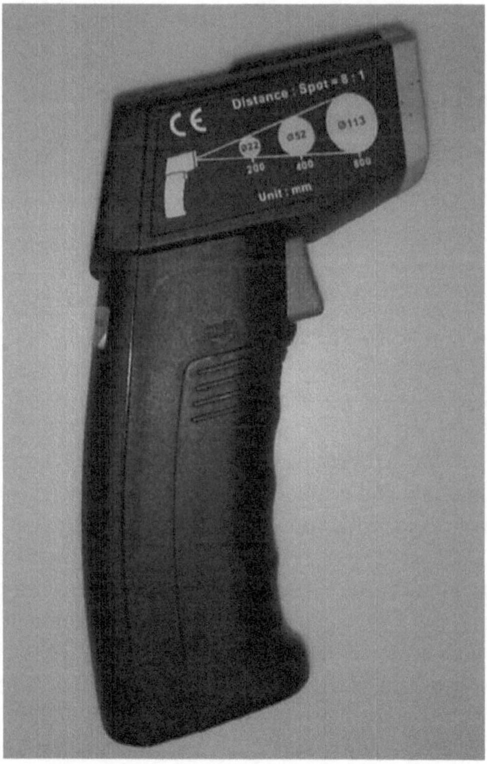

FIGURE 6.23 Probe temperature metre.

an average temperature of the detected region. The temperature reading updated on LCD is noted down for inside and outside of the walls in north, south, east and west directions and measured at three points on the wall, i.e. 1 foot below the roof, at centre and 1 foot above the floor.

FIGURE 6.24 Temperature metre calibration with reference to the temperature of human body and water.

TABLE 6.18
General Specifications of Infrared Thermometer

Display	**:**	**4 digit LCD**
Field of view	:	8:1
Target indicator	:	Laser spot
Emissivity	:	0.98
Temperature range, °C	:	–20–500 (–4–932°F)
Accuracy	:	±2% of reading
Resolution, °F	:	0.5°C/0.5
Repeatability, °C	:	Within ±1% of reading or ±1 (2°F)

Emissivity is a term used to describe the energy-emitting characteristics of materials. Most organic materials and painted or oxidized surfaces have an emissivity of 0.98, and the details of infrared thermometer are shown in Table 6.18. Metal surfaces or shiny materials have a lower emissivity (www.primusthai.com).

Probe temperature meter is the digital temperature meter used to accurately measure internal temperature (i.e. room temperature) and ambient temperature (i.e. atmospheric temperature). Probe temperature meter will ON by pressing the ON/OFF key; during power off status will turn on the system, vice versa. All symbols will be displayed in LCD for two seconds after power on and then the current temperature will be displayed. The temperature unit °C is present after turn on. During measurement can switch it to °C/ °F key.

The probe is extended to measure the internal temperature of an object. Probe is to be inserted at least 12 mm into the target and then the device will measure the core temperature automatically. The probe comes to equilibrium with the object being measured and the reading in LCD reaches almost steady state. Temperature is noted down periodically and stopped at a point when it is in equilibrium. To get a better accuracy, it is measured for a moment after the reading is almost steady to ensure that

TABLE 6.19
General Specifications of Probe Temperature Meter

Display	:	$^{1}/_{3}$ duty, ½ bias 3×13 LCD
Measurement period, sec	:	1 time/2
Auto power off time, min	:	8
Temperature range, °C	:	–50 to 270 (–58 to 518°F)
Accuracy, °C	:	–50–200;±(reading × 1.5% +1°C) 200–270;±(reading × 2.0% +4°C)
Resolution, °C	:	0.1 (–50–200) 1 (200–270)

the probe has fully stabilized and reached equilibrium with the object. Auto power off time is set to 8 min, it will enter into sleep status if there is no action to any key in 8 min after power on. The set time will be re-calculated if any key is pressed. To cancel, the auto power off function °C/°F key is pressed to power on. The specifications of probe temperature metre are mentioned in Table 6.19.

Two rooms are constructed using IOTs and perlite brick and other of ordinary brick (Figure 6.25). After noting the ambient temperature, the temperature of walls on all four sides north wall (NW), south wall (SW), east wall (EW) and west wall (WW) are measured before plastering. The temperature of the roof is noted down.

The two model rooms, i.e. IOT–perlite brick and other of ordinary brick are plastered (Figure 6.26). To the IOT–perlite brick room, mortar used for plastering of wall is of the same proportion of raw materials which is used for preparing IOT–perlite bricks. To the ordinary brick room, cement mortar of 1:4 (Cement: sand) is used for the plastering. The process of noting the temperature of both rooms is done similar to the measurement of temperature before plastering of two model rooms.

A pilot-scale study is taken up to assess the effectiveness of IOT–perlite bricks in actual filed conditions but in a mini scale. Two model brick rooms are constructed and the internal and external temperatures are measured to determine the thermal conductivity of IOT–perlite brick room by comparing the ordinary brick room.

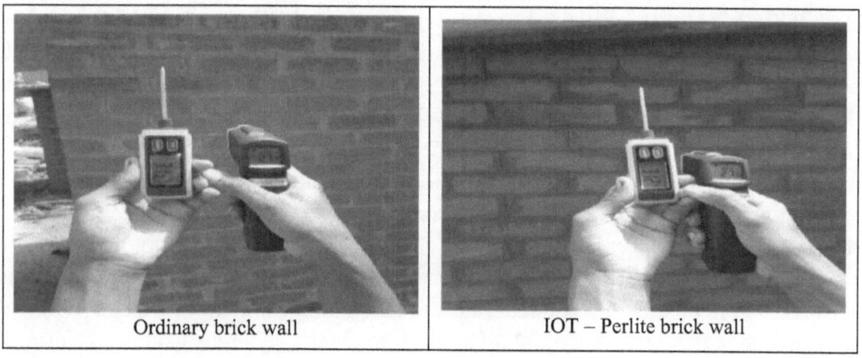

FIGURE 6.25 Temperature measurements of atmosphere and outside wall.

FIGURE 6.26 Model rooms of ordinary and IOT–perlite brick room after plastering.

Model brick rooms are tested before plastering and after plastering. Plaster with IOT and perlite presented better performance due to their low thermal conductivity. These are cost effective and sustainable solutions to passively improve the thermal performance of buildings, as well as to mitigate the impacts of disposal of these residues.

The use of IOT and perlite in bricks reduces the thermal conductivity of the outside wall as these materials arrest the atmospheric temperature from outside compared to the ordinary wall.

One of the trial results of room temperature taken for morning (8 am to 10 am), midday (12 pm to 3 pm) and evening (4 pm to 6 pm) at a regular interval of 15 minutes is given in the following section.

6.4.3 Measurement of Temperature from 8 am to 10 am Before Plastering

Temperature readings are recorded for two model rooms, one with IOT–perlite mix bricks and other with ordinary bricks, at 8 am to10 am on four walls, one set of reading is shown in Table 6.20.

The temperature reading of the outside walls of the IOT–perlite brick and ordinary bricks is taken in different phases during 8 am to 10 am by taking the readings 1 foot below the roof level (Figure 6.27), at middle of the wall (Figure 6.28) and at 1 foot above the floor of the wall (Figure 6.29) in different directions: NW: North side Wall, EW: East side Wall, WW: West side Wall, SW: South side Wall). Similarly, the variation in temperature inside walls of the IOT–perlite bricks and ordinary bricks are shown in Figures 6.30 through 6.32. Temperature is measured through infrared thermometer device. The below-given graphical representation shows the variation in temperature for the walls constructed from IOT mix perlite bricks and ordinary bricks.

From 8 am to 10 am ambient temperature at the surroundings of the rooms is measured and it was 29 °C. The readings are taken on the walls of all sides (N, E, W and S) to check the heat transfer at all sides of the walls, whereas heat is the form of energy transfer from high-temperature location to a low-temperature location.

TABLE 6.20
Temperature Measured from 8 am to 10 am for IOT–Perlite Brick Wall and Ordinary Brick Wall

	IOT–Perlite Bricks								Ordinary Bricks								
	Outside Wall, °C			Inside Wall, °C					Outside Wall, °C			Inside Wall, °C					
	T	M	B	T	M	B	Date	Time	T	M	B	T	M	B	Date	Time	
NW	25.50	25.50	25.00	22.00	21.50	21.50	14–06–2019	10:07 am	29.50	29.50	29.00	26.50	26.00	25.00	14–06–2019	10:00 am	Temperature measured keeping the door closed
EW	26.00	26.00	25.50	22.50	21.50	21.00			30.00	30.00	29.50	27.00	26.50	26.00			
SW	25.50	25.50	25.00	21.50	21.00	20.50			29.50	29.00	28.00	26.00	25.50	25.00			
WW	25.00	24.50	24.00	21.00	20.50	20.00			29.00	28.50	28.00	25.50	25.00	24.50			
Roof	b/w 29.50 & 31.00								Roof			b/w 30.50 & 32.00					
Room Temperature	25.00								Room Temperature			27.50					

Atmospheric Temperature surrounding the constructed rooms – 29.00°C

T – Top (Temperature measured 1 foot below the top)
M – Middle (Temperature measured at centre of the wall)
B – Bottom (Temperature measured 1foot above the floor)

NW – North Wall
EW – East Wall
SW – South Wall
WW – West Wall

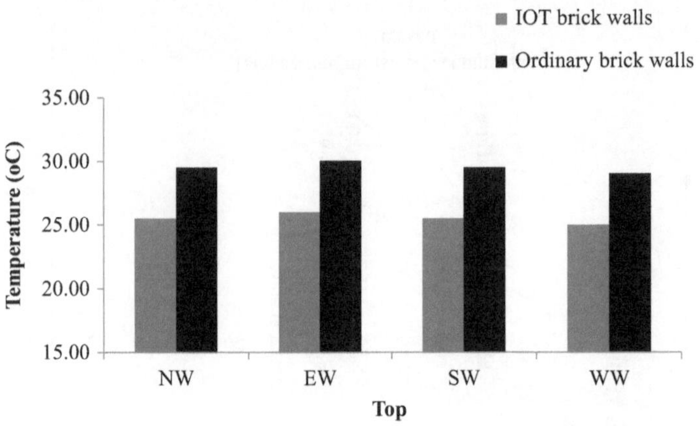

FIGURE 6.27 Temperature measured in IOT–perlite bricks and ordinary bricks of the outside walls at 1 foot below the roof from 8 am to 10 am.

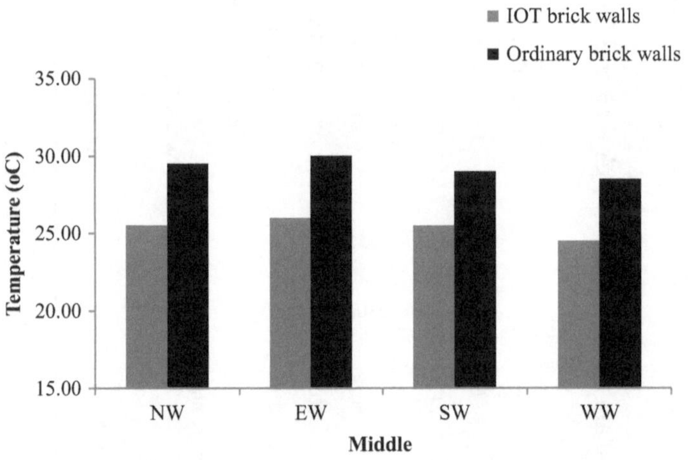

FIGURE 6.28 Temperature measured from 8 am to 10 am at middle of the outside walls.

The sun rays fall on eastern side of the wall during forenoon, so the temperature outside of east wall is more compared to other sides. On an average, IOT–perlite bricks walls resulted in 3–4 °C less temperature than ordinary bricks.

It is also observed that temperature of the IOT–perlite brick walls in outer surface is less by 10% than ambient temperature, which is only 0%–2% in the case of ordinary brick walls. The temperature of IOT–perlite brick walls from outer to inner surface is reduced by 13%, but in case of ordinary bricks it is 10%. This concludes that the ambient temperature is arrested by outer surface of the IOT–perlite brick walls by more than 10% and inner surface is at least 3% less compared to that of ordinary brick walls. The room temperature of IOT–perlite bricks is less by 7% compared to that of ordinary bricks.

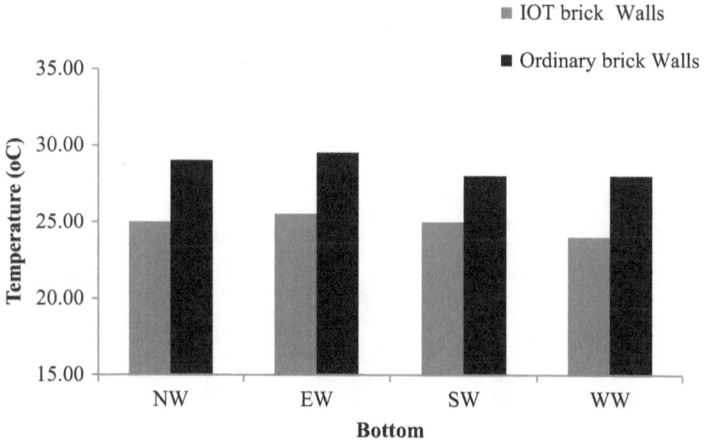

FIGURE 6.29 Temperature measured at 1 foot above the floor of outside walls from 8 am to 10 am.

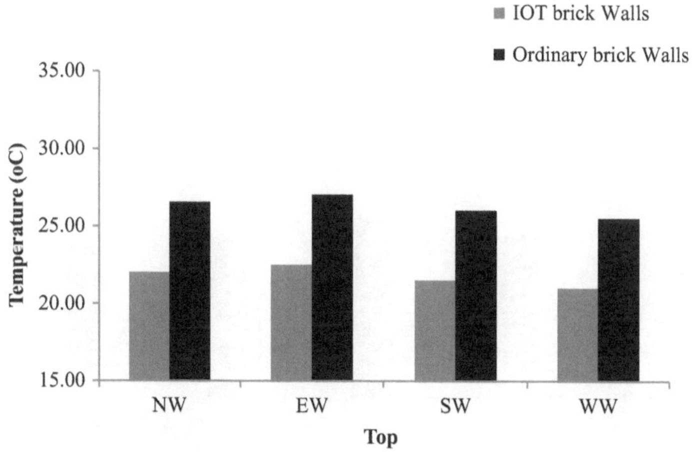

FIGURE 6.30 Temperature of the inside walls measured at 1 foot below the roof from 8 am to 10 am.

6.4.4 Measurement of Temperature from 12 pm to 3 pm Before Plastering

For the constructed two model rooms, one with IOT and perlite mix bricks and other with ordinary bricks, following are the temperature readings observed from 12 pm to 3 pm for four side walls (Table 6.21). Six temperature measurements are taken for midday, in which one of the measurements is mentioned in Table 6.21.

The temperature readings taken on inside and outside walls of IOT–perlite bricks and ordinary brick walls during 12 pm to 3 pm plotted to know the difference in

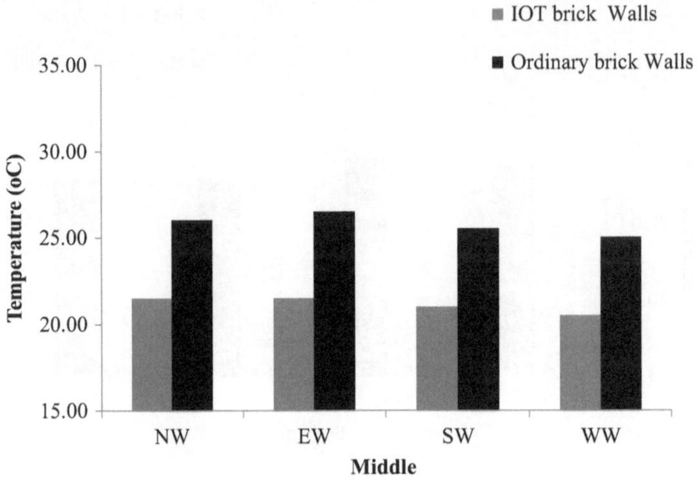

FIGURE 6.31 Temperature measured from 8 am to 10 am at middle of the inside walls.

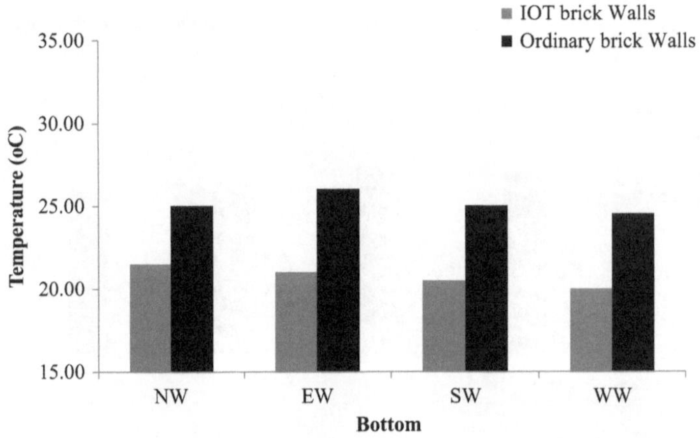

FIGURE 6.32 Temperature measured from 8 am to 10 am at 1 foot above the floor of inside walls.

temperature occurred. Figure 6.33 indicates the comparison of temperature when it is measured 1 foot below the roof of the outside walls; Figure 6.34 shows the temperature at middle of the walls at outside walls. Temperature variation of the outside walls at 1 foot above the floor is represented in Figure 6.35. The inside walls temperature difference is represented in Figures 6.36 through 6.38.

The radiation from outside to inside walls is transferred by process of conduction through all directions which are seldom steady. The temperature at midday is higher at the east, north and south walls compared to the west wall. Ambient temperature

TABLE 6.21
Temperature Measured from 12 pm to 3 pm for Outside and Inside of IOT–Perlite Bricks and Ordinary Bricks Wall

	IOT Bricks								Ordinary Bricks								
	Outside Wall, °C			Inside Wall, °C					Outside Wall, °C			Inside Wall, °C					
	T	M	B	T	M	B	Date	Time	T	M	B	T	M	B	Date	Time	
NW	28.50	29.00	27.50	25.50	24.50	23.50			32.50	31.50	31.00	28.00	26.50	26.00			Temperature measured keeping the door closed
EW	28.50	28.00	28.50	26.00	25.00	24.00			32.00	32.00	31.50	28.00	26.50	26.00			
SW	29.00	28.50	27.00	25.00	24.50	23.00	29–06–2019	12:08 pm	32.50	31.50	31.50	27.50	26.00	25.50	29–06–2019	12:00 pm	
WW	28.00	27.50	25.50	25.50	24.50	23.00			30.00	31.00	30.00	27.00	26.00	25.00			
Roof					b/w 38.00 & 43.50					Roof			b/w 36.50&39.00				
Room Temperature					27.50					Room Temperature			29.00				

Atmospheric Temperature surrounding the constructed rooms – 30.00°C

T – Top (Temperature measured 1 foot below the top)
M – Middle (Temperature measured at centre of the wall)
B – Bottom (Temperature measured 1foot above the floor)

NW – North Wall
EW – East Wall
SW – South Wall
WW – West Wall

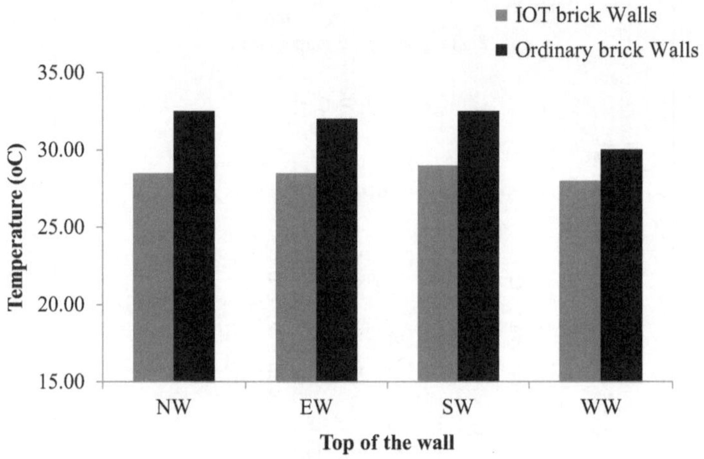

FIGURE 6.33 Temperature measured from 12 pm to 3 pm at 1 foot below the roof of the outside walls.

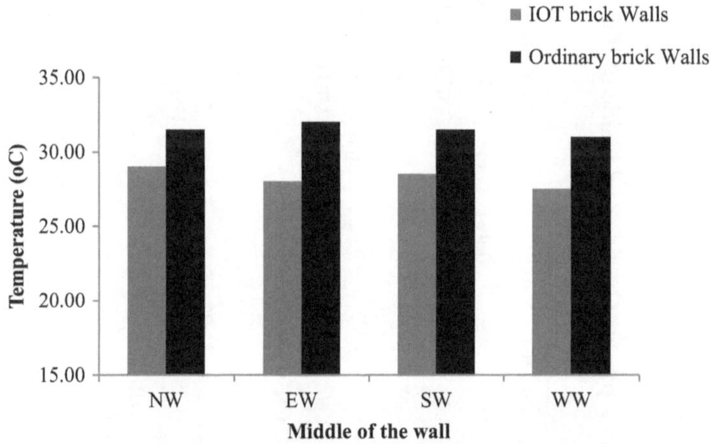

FIGURE 6.34 Temperature measured from12 pm to 3 pm at middle of the outside walls.

arrested by IOT–perlite brick from outside surface is 5% more when compared to ordinary bricks. The transfer of heat from outside to inside of the ordinary and IOT–perlite brick is less by 13% and 14%, respectively; ordinary bricks have shown 1% less temperature difference. But, by and large when the temperature of inside surface of the bricks is measured, IOT–perlite bricks are found slightly less compared to that ordinary bricks. It justifies the lower effect of radiation and conduction of the bricks made of IOT and Perlite. The room temperature of IOT–perlite brick room is 5% less than the ordinary brick room. It could be concluded that, in midday, the temperature at walls, floor and room of IOT–perlite bricks is low compared to conventional bricks because of thermal resistivity of perlite.

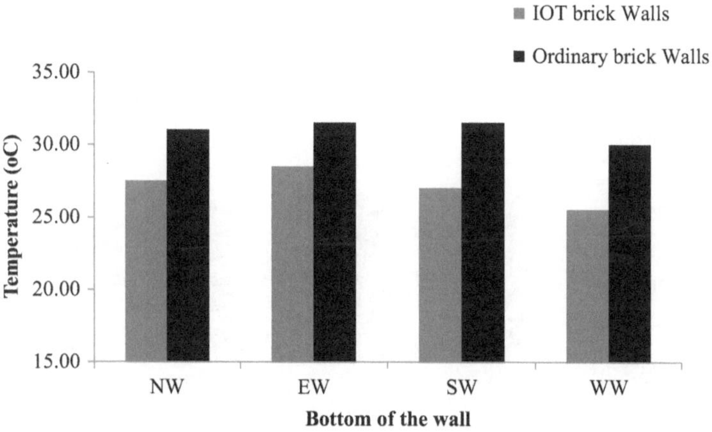

FIGURE 6.35 Temperature measured from 12 pm to 3 pm at 1 foot above the floor of the outside walls.

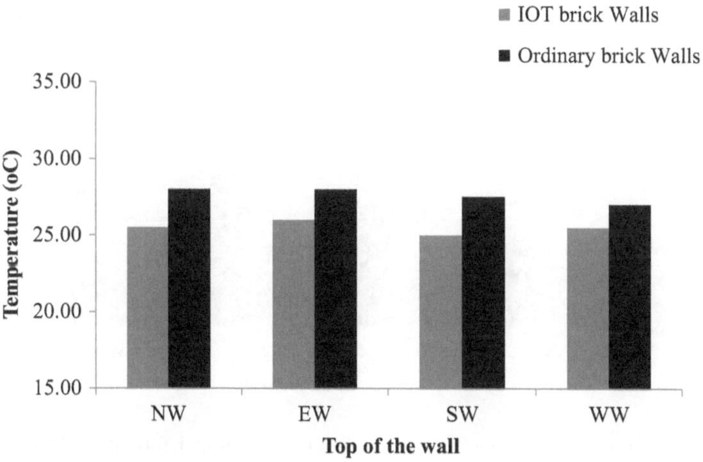

FIGURE 6.36 Temperature measured from 12 pm to 3 pm at 1 foot below the roof of the inside walls.

6.4.5 Measurement of Temperature from 4 pm to 6 pm Before Plastering

For the constructed two model rooms, one with IOT and perlite mix bricks and other with ordinary bricks, following are the temperature readings observed from 4 pm to 6 pm for four side walls (Table 6.22). The temperature of IOT–perlite and ordinary bricks is taken for inside and outside walls, room and atmospheric temperature in 6 nos, in which one is mentioned in Table 6.22.

Similar to morning and afternoon, the temperature readings in the evening are also taken for side of the brick walls in all directions. It is to know the ambient temperature arrested in IOT–perlite brick and ordinary brick which are shown in Figures 6.39

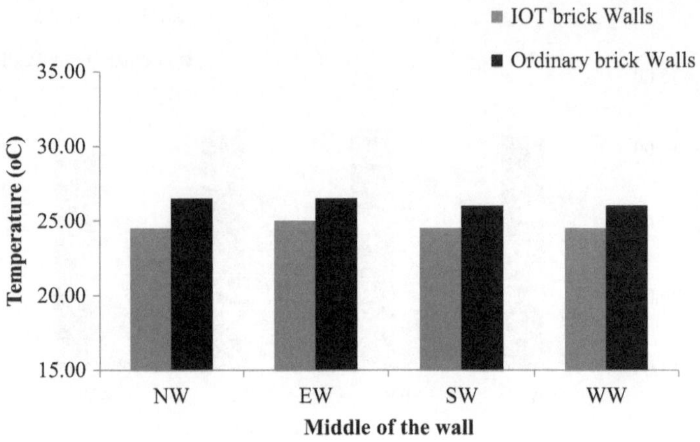

FIGURE 6.37 Temperature measured from 12 pm to 3 pm at middle of the inside walls.

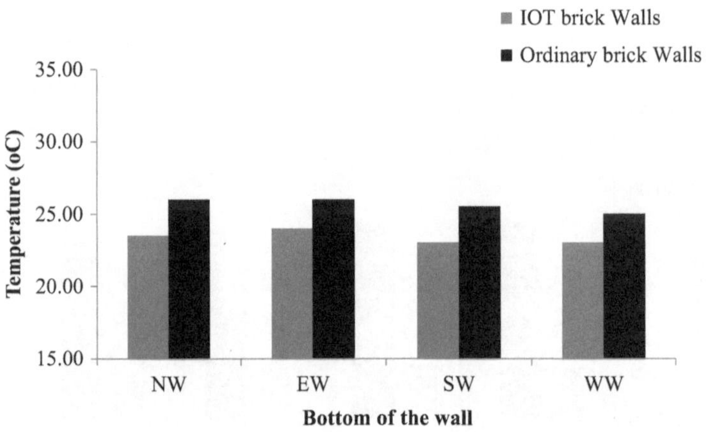

FIGURE 6.38 Temperature measured from 12 pm to 3 pm at 1 foot above the floor of the inside walls.

through 6.41 for 1 foot below roof, at the middle of wall and 1 foot above the ground, respectively. The temperature of inner surface of the walls in all directions of both the model rooms are measured and plotted in Figures 6.42 through 6.44.

The atmospheric temperature in the evening is noted as 28°C. The temperatures in each wall of the rooms (IOT–perlite bricks and ordinary bricks) are not same and it is varied by at least 3% to 4%. Temperature of the IOT–perlite brick walls in outer surface is less by 3% than ambient temperature, which is only 0%–1% in the case of ordinary brick walls. The radiation of the sunlight conducted in the IOT–perlite brick walls from outer to inner surface is less by 4%, but in case of ordinary bricks, 2% temperature is lowered. This concludes that the ambient temperature arrested by outer surface of the IOT–perlite brick walls is more than 2% and inner surface is at

TABLE 6.22
Temperature Measured from 4 pm to 6 pm on IOT–Perlite Brick Wall and Ordinary Brick Wall

	IOT Bricks								Ordinary Bricks								
	Outside Wall, °C			Inside Wall, °C					Outside Wall, °C			Inside Wall, °C					
	T	M	B	T	M	B	Date	Time	T	M	B	T	M	B	Date	Time	
NW	27.50	27.50	27.00	25.50	26.50	26.00			30.50	30.50	29.50	29.00	29.50	28.00			
EW	26.00	26.50	26.50	25.00	26.00	25.00			30.00	29.50	28.00	29.00	-29.00	27.50			
SW	27.50	27.50	26.50	26.00	26.00	26.00	15–06–2019	6:05 pm	31.00	30.00	27.50	30.50	29.50	27.00	15–06–2019	5:55 pm	Temperature measured keeping the door closed
WW	27.50	27.50	27.00	26.00	26.50	26.00			31.50	30.50	30.00	31.00	30.00	29.00			
Roof				b/w 28.00 & 29.00					Roof			b/w 28.50 & 30.50					
Room Temperature				26.50					Room Temperature			28.00					

Atmospheric Temperature surrounding the constructed rooms – 28°C

T – Top (Temperature measured 1 foot below the top)
M – Middle (Temperature measured at centre of the wall)
B – Bottom (Temperature measured 1foot above the floor)

NW – North Wall
EW – East Wall
SW – South Wall
WW – West Wall

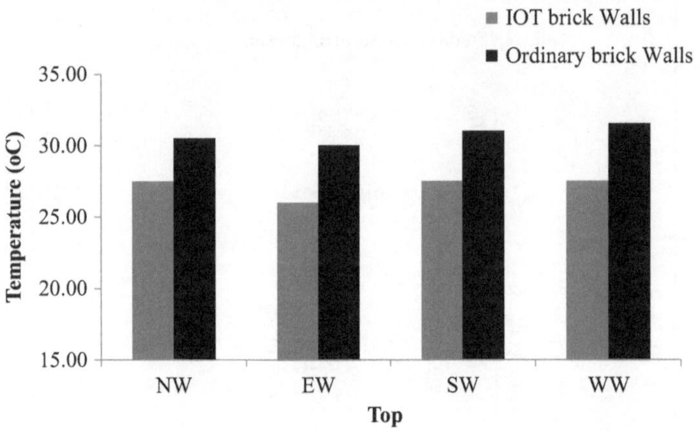

FIGURE 6.39 Temperature measured from 4 pm to 6 pm of the outside walls at 1 foot below the roof,

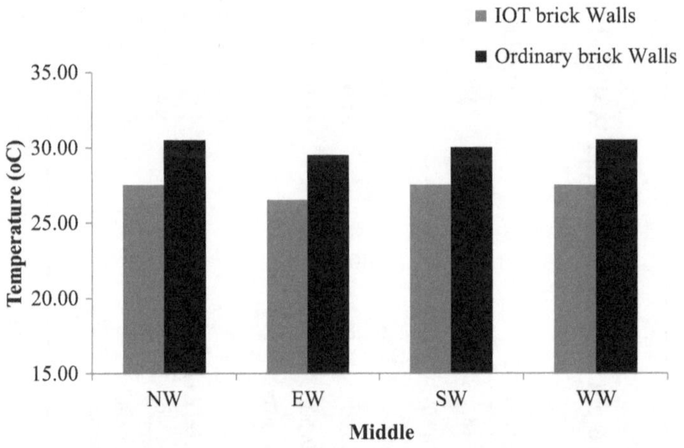

FIGURE 6.40 Temperature measured from 4 pm to 6 pm of the outside walls at middle of the wall.

least 2% less compared to that of ordinary brick walls. The inside room temperature of IOT–perlite bricks is less by 5% compared to that of ordinary bricks. It is concluded that the temperature is varied in all walls of the four sides, but as an average IOT–perlite bricks room results in 1 to 2°C less in temperature than ordinary bricks.

In all the above cases, at each location 8 to 10 readings were taken and only average values are reported. The wall of IOTs and perlite brick room has given the better result in all the cases.

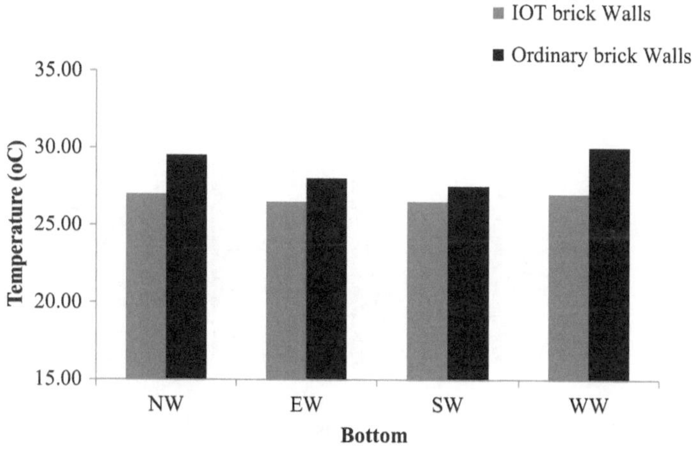

FIGURE 6.41 Temperature measured from 4 pm to 6 pm of the outside walls at 1 foot above the floor.

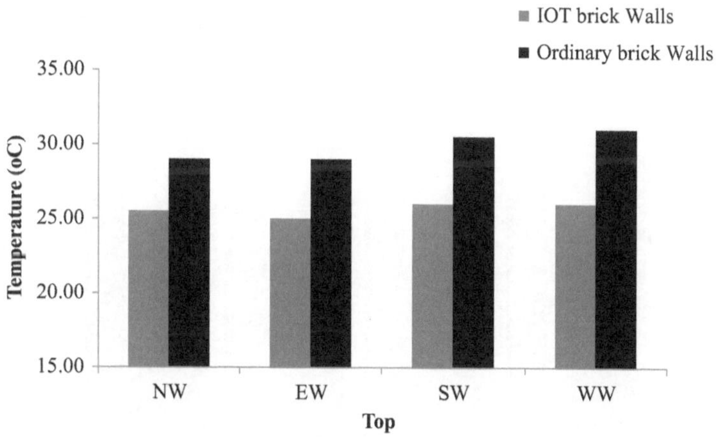

FIGURE 6.42 Temperature measured from 4 pm to 6 pm of the inside walls at 1 foot below the roof.

6.4.6 Temperature Measurement of the Room at 5 Minutes Interval Before Plastering

Temperature is a physical property of matter that quantitatively expresses common notions of hot and cold. Room temperature and ambient temperature are directly affected by the heating and cooling system. The solar energy received by any region varies with time of day, with seasons and with latitude that causes temperature variations. Hence, influence of time on room temperature is determined to know the thermal behaviour causing the change in IOT–perlite brick room, and the readings of

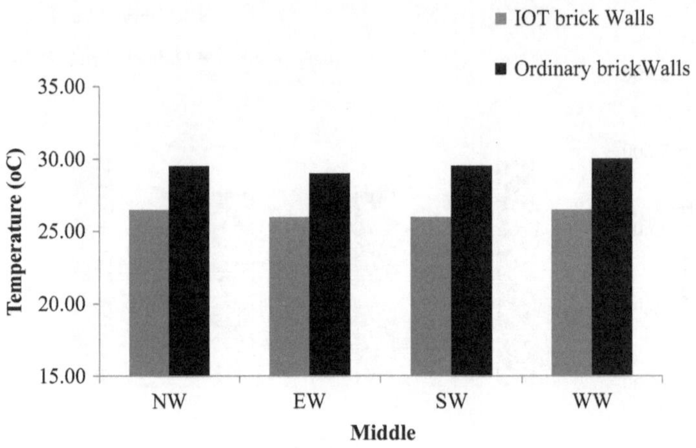

FIGURE 6.43 Temperature measured from 4 pm to 6 pm of the inside walls at middle of the wall.

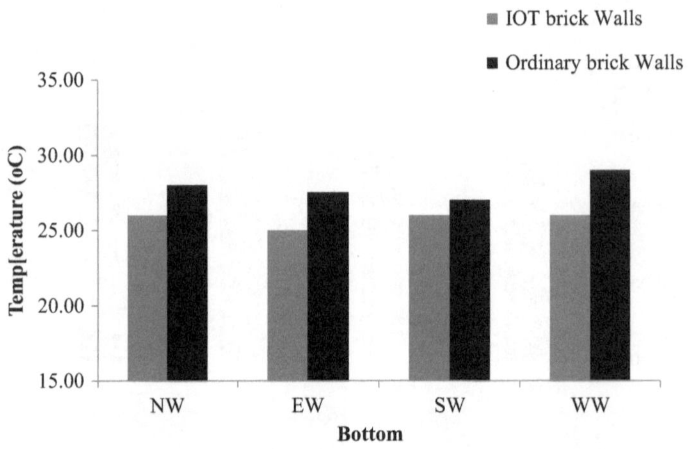

FIGURE 6.44 Temperature measured from 4 pm to 6 pm of the inside walls at 1 foot above the floor.

IOT–perlite brick room and ordinary brick room is noted for every 5 minutes interval (Table 6.23).

The two rooms are plastered with 12 mm thick mortar on both sides of the wall: the room of ordinary brick room with cement mortar and IOTs and perlite brick room with the mortar same as of material and proportion of that brick. The quality of mortar obtained in IOTs and perlite brick wall plastering is same as of the quality of its brick. The room temperature is measured with probe temperature metre by closing the door. The comparison of temperature difference is done with ambient temperature, IOTs

TABLE 6.23
Temperature Measurement at 5 Minutes Interval Before Plastering

Time	Ambient Temperature, °C	Ordinary Brick Room, °C	IOT–Perlite brick room, °C
0	35.80	35.70	34.60
5	34.50	34.60	33.50
10	35.00	35.00	33.90
15	35.10	35.50	34.20
20	36.00	36.00	34.30
25	35.00	35.70	34.60
30	35.60	35.80	34.50
35	35.60	35.60	34.20
40	35.60	35.80	34.00
45	35.10	35.70	34.40
50	35.50	36.10	34.70
55	35.30	36.70	35.65

and perlite brick room and ordinary brick room. The test is done after the plastering to ensure further reduction in room temperature and to obtain less heat and minimize the energy consumption in air conditioning system.

To confirm the thermal efficiency of model room made of bricks with mine waste such as IOT and perlite, the room temperature is checked for longer duration. The temperature variation in IOT–perlite brick room and ordinary brick room are measured at a regular interval of 5 minutes for 55 minutes. It is noticed that the room temperature varies with a change in atmospheric temperature but maintains the average difference at all points of time. It is established that the temperature of the IOT–perlite brick room is less by 1–2°C compared to an ordinary brick room (Figure 6.45). This helps to prove that the radiation, conduction and convection of

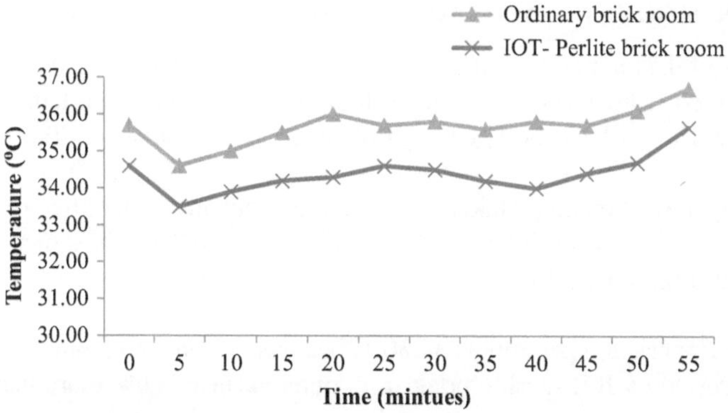

FIGURE 6.45 Temperature vs time before plastering of model rooms.

heat through IOT–perlite brick is less compared to ordinary brick room by maintaining difference in temperature proportionately throughout with the variation in ambient temperature.

6.4.7 Measurement of Temperature from 8 am to 10 am After Plastering

The same procedure is followed to measure the temperature of outside of the walls, inside of the walls and room of IOT–perlite brick and ordinary brick. Here one set of the reading is presented in Table 6.24.

The temperature of ordinary brick room as well as the IOTs and perlite brick room is affected by ambient temperature and solar radiation. In peak condition also temperature of IOTs and perlite brick room is less than ordinary brick room and ambient temperature. This decrease in wall and room temperature caused in IOTs and perlite brick room is due to addition of perlite in brick, where perlite acts as good thermal-resistant material. The perlite arrests the ambient temperature and solar radiation from outside and releases less heat into the room, which helps to keep the room cool. The comparison of IOT–perlite brick wall and ordinary brick wall temperature is expressed in graphical representation as shown in the Figures 6.46 through 6.48 for outside walls and Figures 6.49 through 6.51 for inside walls.

When the ambient temperature was 30°C, the result concludes that after wall plastering the temperature of the IOT–perlite brick walls in outer surface reduced by 12% than ambient, but in ordinary brick walls temperature arrested 2% less compared to IOT–perlite brick walls. This difference is compared to without plastering IOT–perlite walls and observed 2% further reduction after plastering of IOT–perlite walls. The room temperature of IOT–perlite brick is less by 8% compared to ordinary brick. When it is compared with without plastering, 1% further reduction was obtained. Therefore, it gives the better result after plastering of IOTs and perlite bricks room and also can save the energy consumption in the room because of lower temperature.

6.4.8 Measurement of Temperature from 12 pm to 3 pm After Plastering

Several sets of the temperature of walls of IOT–perlite brick room and ordinary brick room, observed from 12 pm to 3 pm for four side walls and room temperature, are measured and one of the best readings obtained is mentioned in Table 6.25 as a sample.

The temperature readings taken on inside and outside walls after plastering of IOT–perlite bricks and ordinary bricks during 12 pm to 3 pm is shown in Figures 6.52 through Figure 6.54 for outside walls and Figures 6.55 through 6.57 for inside walls.

The atmospheric temperature is 34.50°C between 12 pm and 3 pm. Ambient temperature arrested by IOT–perlite brick from outer surface is 6% more than ordinary bricks and 1% more than non-plastered IOT–perlite brick wall. The transfer of heat from outside to inside of the ordinary and IOT–perlite brick is less by 5% and 7%, respectively; ordinary brick have shown 2% less temperature difference. The room

TABLE 6.24
Temperature Measured from 8am to 10am on IOT–Perlite Brick Wall and Ordinary Brick Wall

	IOT Bricks								Ordinary Bricks							
	Outside Wall, °C			Inside Wall. °C					Outside Wall, °C			Inside Wall, °C				
	T	M	B	T	M	B	Date	Time	T	M	B	T	M	B	Date	Time
NW	23.50	23.50	22.50	23.50	22.00	20.50	24–09–2019	10:30 am	24.50	24.00	23.00	24.00	23.00	22.50	24–09–2019	10:40 am
EW	25.00	24.50	23.50	24.50	23.50	22.00			26.00	25.50	24.00	26.00	25.00	23.50		
SW	30.00	31.00	31.50	28.00	27.00	25.50			30.50	32.00	32.50	29.50	29.50	28.00		
WW	28.00	27.50	26.00	25.50	24.50	22.50			29.00	28.50	28.00	28.00	27.00	26.00		
Roof				30.50 °C – 38.00 °C					Roof			35.50 °C – 38.50 °C				
Room Temperature				25.50 °C					Room Temperature			28.00 °C				

Temperature measured keeping the door closed

Atmospheric Temperature surrounding the constructed rooms – 30.00°C

T – Top (Temperature measured 1 foot below the top)
M – Middle (Temperature measured at centre of the wall)
B – Bottom (Temperature measured 1foot above the floor)

NW – North Wall
EW – East Wall
SW – South Wall
WW – West Wall

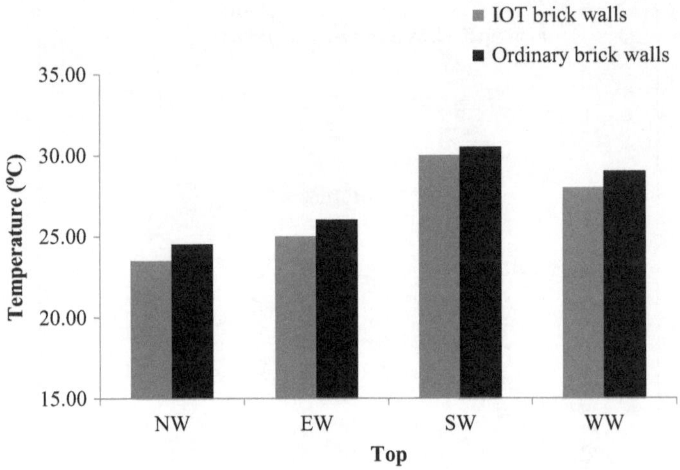

FIGURE 6.46 Temperature measured from 8 am to 10 am at top point of the outside walls.

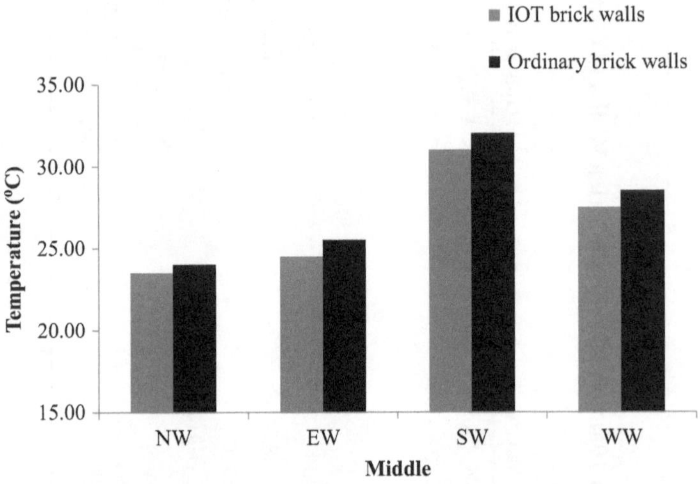

FIGURE 6.47 Temperature measured from 8 am to 10 am at middle point of the outside walls.

temperature of IOT–perlite brick is less by 8% compared to ordinary brick. When it is compared with without plastering 1% further reduction was observed.

6.4.9 Measurement of Temperature from 4 pm to 6 pm After Plastering

The temperature readings of IOT–perlite brick room and ordinary brick room, observed from 4 pm to 6 pm for four side walls and room temperature are given in Table 6.26 as sample data.

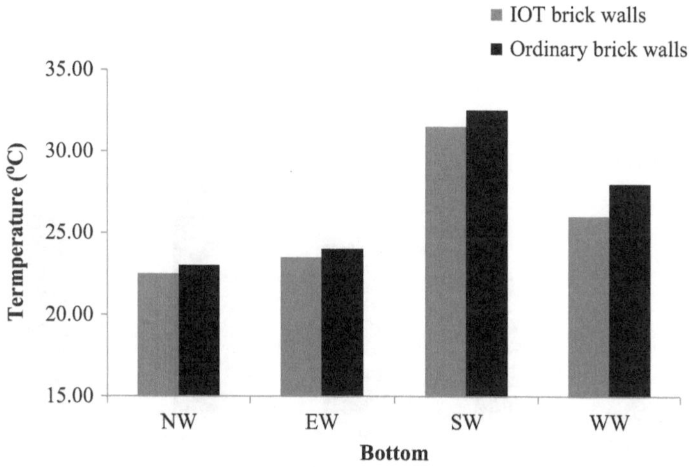

FIGURE 6.48 Temperature measured from 8 am to 10 am at bottom point of the outside walls.

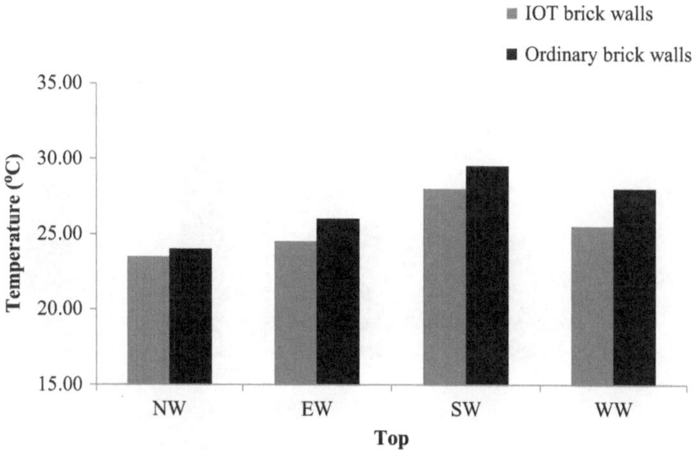

FIGURE 6.49 Temperature measured from 8 am to 10 am at top point of the inside walls.

Similar to morning and afternoon, the temperature readings in the evening is noted. Figures 6.58 through 6.60 for top, middle and bottom, respectively, for outside wall of IOT–perlite brick and ordinary brick temperature is measured. The inside wall temperature of the IOT–perlite brick and ordinary brick is noted and shown in Figures 6.61 through 6.63.

The ambient temperature in the evening was noted as 34.00 °C. The temperature of IOT–perlite brick walls in outer surface is less by 14%, it says that the temperature is well arrested at outer wall of the IOT–perlite brick wall. But in case of ordinary

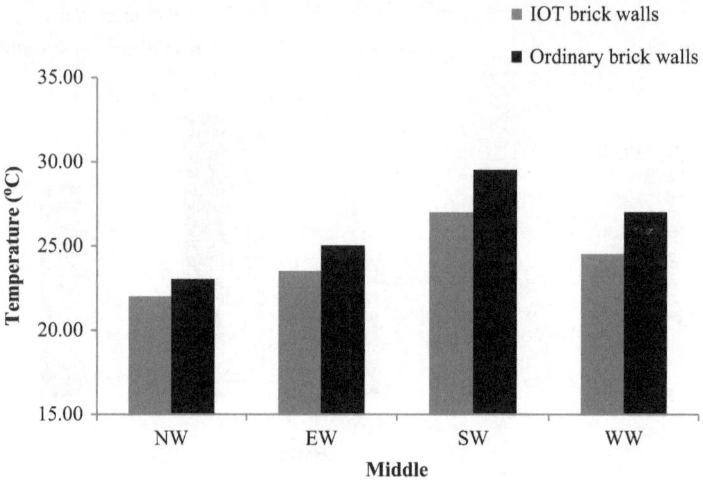

FIGURE 6.50 Temperature measured from 8 am to 10 am at middle point of the inside walls.

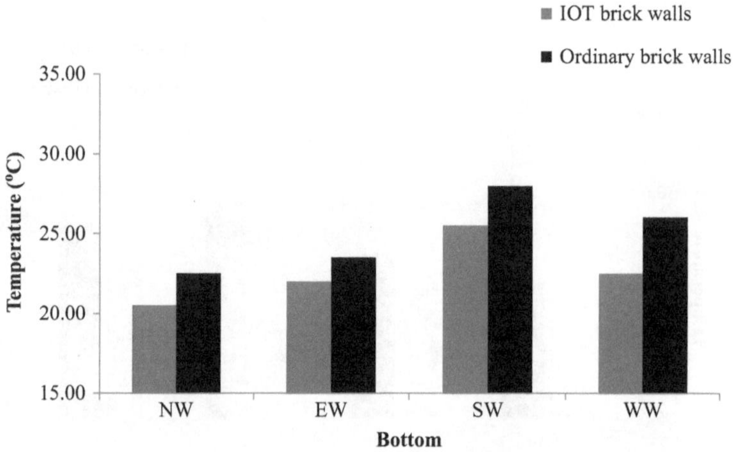

FIGURE 6.51 Temperature measured from 8 am to 10 am at bottom point of the inside walls.

brick wall it arrests about only 8%; hence, it proves that IOT–perlite brick walls reduce the temperature. The room temperature of IOT–perlite brick is less by 8% compared to ordinary brick. When it is compared with without plastering 1% further reduction was obtained.

6.4.10 Temperature Measurement of the Room at 5 Minutes Interval After Plastering

Temperature measured with reference to time in an interval of 5 minutes for 55 minutes, after free ventilation was closed (Table 6.27). Influence of time on room

TABLE 6.25
Temperature Measured from 12pm to 3pm on IOT–Perlite Brick Wall and Ordinary Brick Wall

	IOT Bricks								Ordinary Bricks								
	Outside Wall, °C			Inside Wall, °C					Outside Wall, °C			Inside Wall, °C					
	T	M	B	T	M	B	Date	Time	T	M	B	T	M	B	Date	Time	
NW	32.50	31.00	28.00	31.00	29.00	26.50	27–09–2019	1:20 pm	33.50	32.00	31.50	32.50	30.50	30.50	27–09–2019	1:25 pm	Temperature measured keeping the door closed
EW	30.00	29.00	28.00	28.50	27.00	27.00			30.50	29.50	29.00	29.50	28.00	28.00			
SW	32.50	33.00	33.50	30.00	31.50	31.00			33.00	33.50	34.50	31.00	32.00	33.50			
WW	30.50	31.00	31.50	28.00	29.50	29.50			31.50	31.50	32.50	30.50	30.00	31.00			
Roof				40.50°C–42.00°C					Roof			41.50°C–43.50°C					
Room Temperature				31.00°C					Room Temperature			33.50°C					

Atmospheric Temperature surrounding the constructed rooms – 34.50°C

T – Top (Temperature measured 1 foot below the top)
M – Middle (Temperature measured at centre of the wall)
B – Bottom (Temperature measured 1foot above the floor)

NW – North Wall
EW – East Wall
SW – South Wall
WW – West Wall

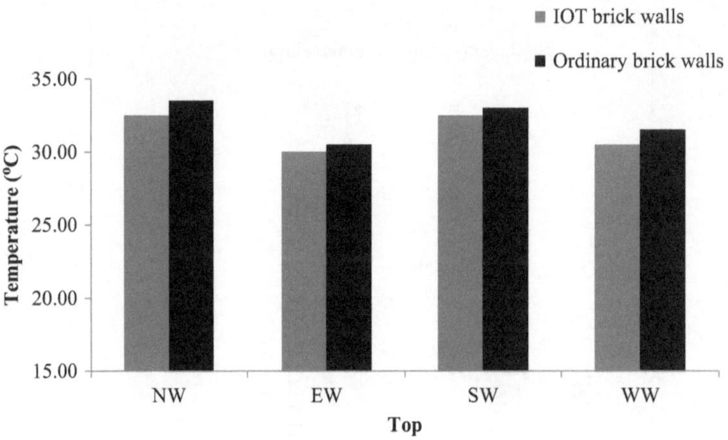

FIGURE 6.52 Temperature measured from 12 pm to 3 pm at top point of the outside walls.

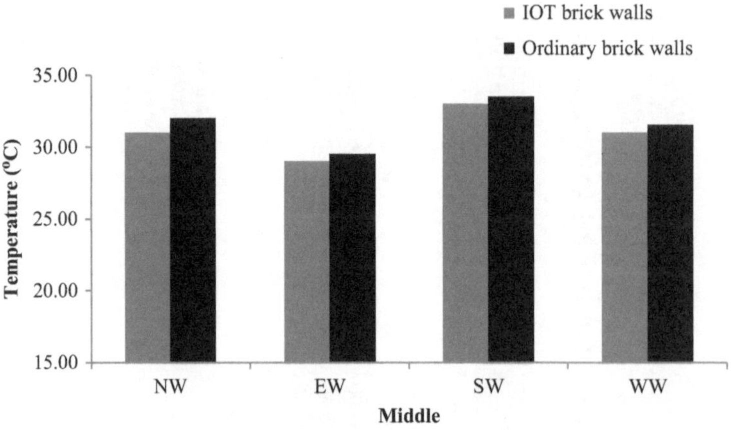

FIGURE 6.53 Temperature measured from 12 pm to 3 pm at middle point of the outside walls.

temperature is determined after plastering of IOT–perlite brick room and ordinary brick room to know the thermal behaviour causing the change in both the rooms.

The temperature variation in IOT–perlite brick room and ordinary brick room after plastering are measured at a regular interval of 5 minutes for 55 minutes. It is noticed that the room temperature varies with a change in atmospheric temperature but maintains the average difference at all points of time. It is established that the temperature of the IOT–perlite brick room is less by 2–2.50°C compared to an ordinary brick room (Figure 6.64). Figure 6.65 shows the temperature response of proto type room for full day. This concludes that using perlite as an additive is given an advantage, which reduced the room temperature and obtained 0.5 °C further reduction compared to without plastering, which saves the energy consumption.

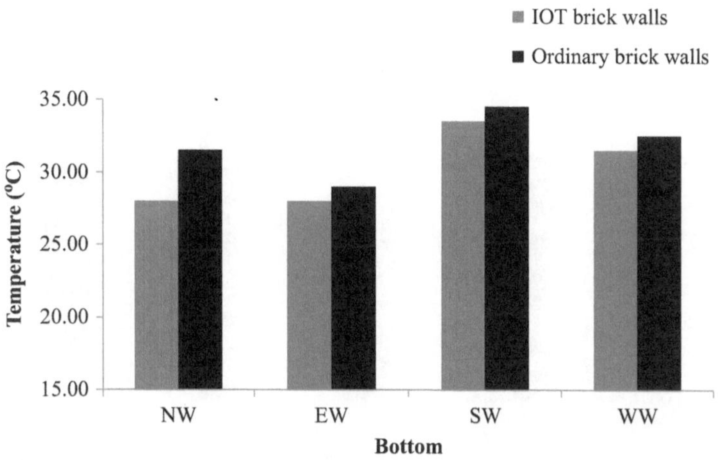

FIGURE 6.54 Temperature measured from 12 pm to 3 pm at bottom point of the outside walls.

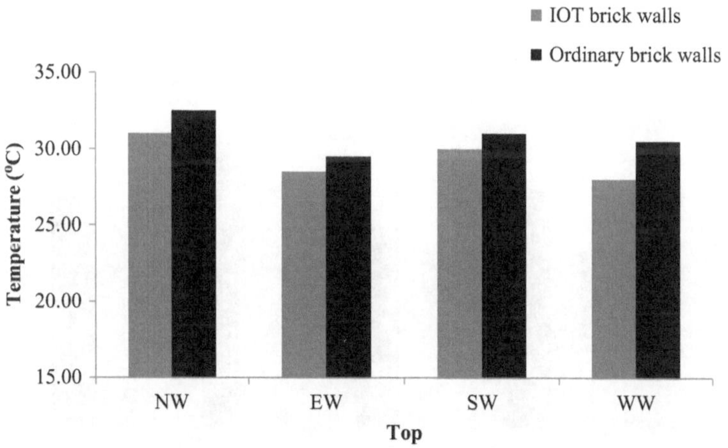

FIGURE 6.55 Temperature measured from 12 pm to 3 pm at top point of the inside walls.

6.4.11 Cost Saving in Terms of Energy Consumption

Percentage of energy or money saved by using IOT–perlite bricks, compared to ordinary bricks is analysed and calculated with reference to the energy required by air conditioner to reduce the room temperature by 2°C, is presented in the Table 6.28. Air conditioner of 1.5 tonnes is considered for the calculation of percentage of saved energy. At atmospheric temperature 31°C, the room temperature of IOT–perlite bricks and ordinary bricks are 25°C and 27°C, respectively, in one of the selected trials and considered the IOT–perlite brick room temperature is less by 2°C compared to ordinary brick room.

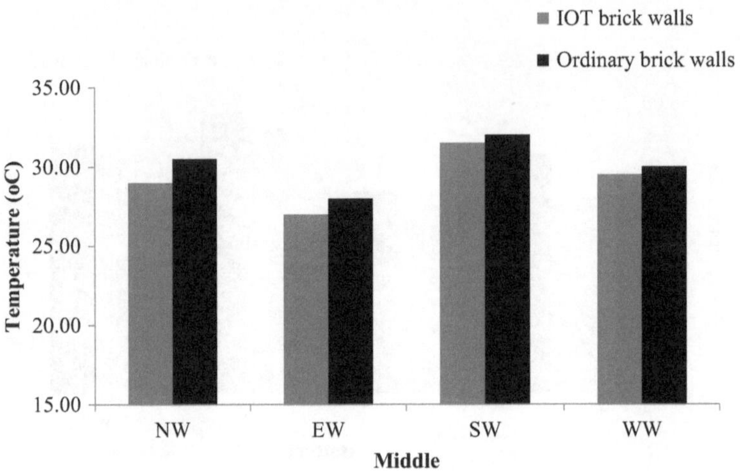

FIGURE 6.56 Temperature measured from 12 pm to 3 pm at middle point of the inside walls.

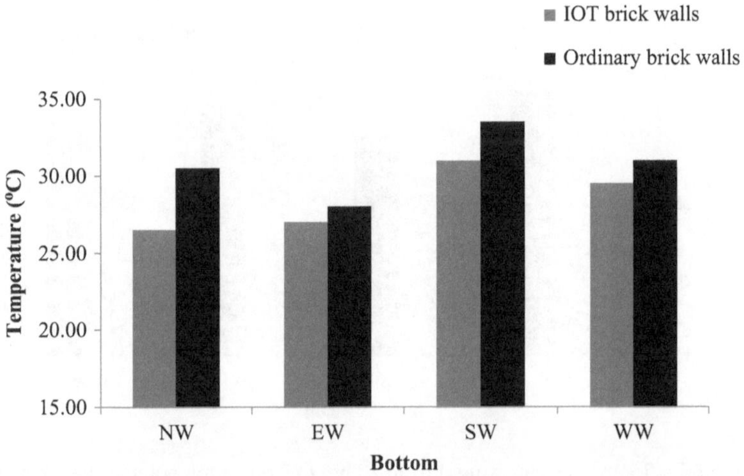

FIGURE 6.57 Temperature measured from 12 pm to 3 pm at bottom point of the inside walls.

As shown in Table 6.28, to maintain room temperature of 18°C as reference temperature, the electricity charges are Rs.2633 per month. Similarly, to maintain 25 °C and 27 °C, the electricity charges are Rs. 2025 and Rs. 1823, respectively. The study has obtained the average room temperature for IOT–perlite brick room, i.e. 25 °C, which does not require air conditioner usually at this point. If the room needs to maintain temperature of 18 °C in IOT–perlite brick room, when the temperature of IOT–perlite brick room is 25 °C the electricity charges will be Rs. 608 per month (Rs.2633–Rs.2025).

TABLE 6.26
Temperature Measured from 4 pm to 6 pm on IOT–Perlite Brick Wall and Ordinary Brick Wall

	IOT Bricks								Ordinary Bricks								
	Outside Wall, °C			Inside Wall, °C					Outside Wall, °C			Inside Wall, °C					
	T	M	B	T	M	B	Date	Time	T	M	B	T	M	B	Date	Time	
NW	30.00	29.50	28.00	27.50	27.00	25.50	30–09–2019	4:10 pm	30.50	30.00	29.00	29.50	29.00	28.00	30–09–2019	4:15 pm	Temperature measured keeping the door closed
EW	30.50	29.00	27.50	28.00	26.50	25.00			30.00	29.00	28.00	29.00	27.50	27.00			
SW	31.50	32.50	33.00	29.00	30.00	30.50			32.50	33.00	33.50	31.50	32.00	32.50			
WW	31.00	32.00	34.00	28.50	29.00	31.50			31.50	33.00	35.00	30.50	32.00	34.00			
Roof				38.00°C–39.00°C					Roof			39.00°C–40.00°C					
Room Temperature				29.00°C					Room Temperature			31.50°C					

Atmospheric Temperature surrounding the constructed rooms – 34.00°C

T – Top (Temperature measured 1 foot below the top)
M – Middle (Temperature measured at centre of the wall)
B – Bottom (Temperature measured 1foot above the floor)

NW – North Wall
EW – East Wall
SW – South Wall
WW – West Wall

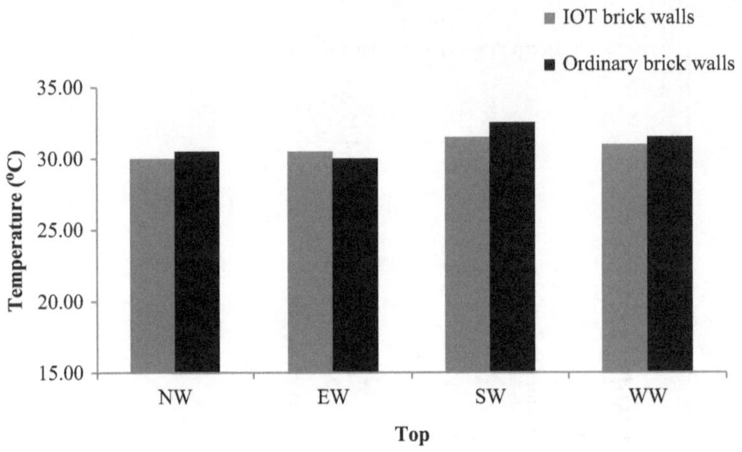

FIGURE 6.58 Temperature measured from 4 pm to 6 pm at top point of the outside walls.

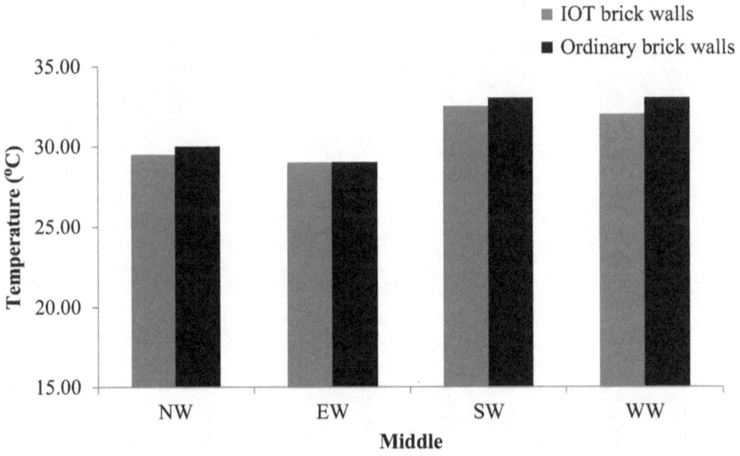

FIGURE 6.59 Temperature measured from 4 pm to 6 pm at middle point of the outside walls.

Accordingly, the room temperature of ordinary brick is 27 °C as an average obtained from the trails, the electricity charges are Rs. 1823 per month. If the room requires 18 °C, when the room temperature of ordinary brick is 27 °C, the electricity charges will be Rs. 810 per month (Rs.2633–Rs.1823).

It concludes that the electricity charges of Rs. 202 per month is more in ordinary bricks room compared to IOT–perlite brick room. Hence, there is saving of 8% energy in the IOT–perlite brick room, in terms of electricity for room size of 48 cubic metre because of lower thermal conductivity in IOT–perlite bricks. The estimations are made as per the ready reckoner of The Energy and Resources Institute, TERI (https://www.livemint.com).

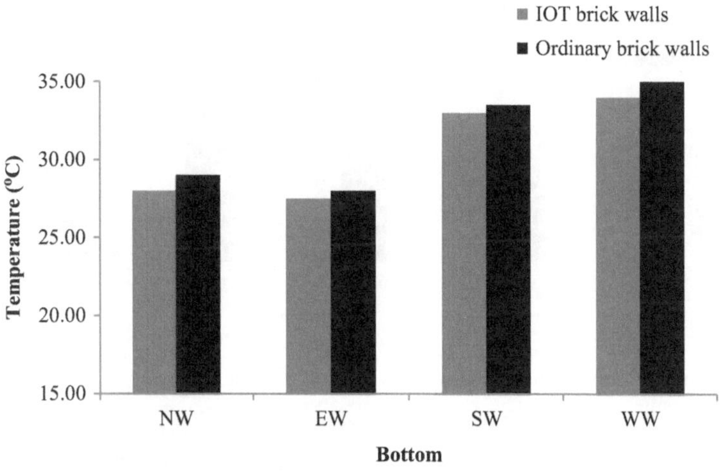

FIGURE 6.60 Temperature measured from 4 pm to 6 pm at bottom point of the outside walls.

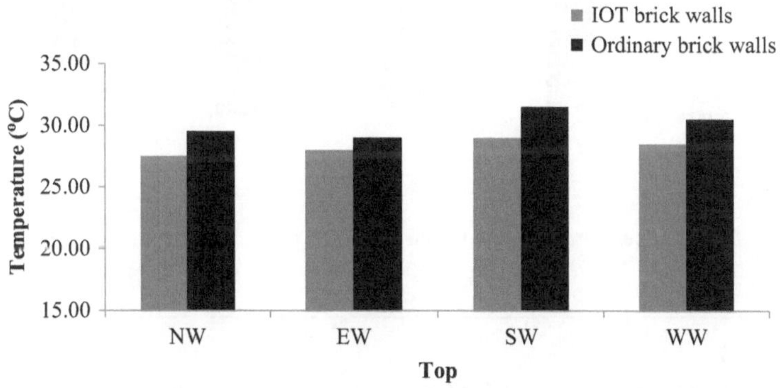

FIGURE 6.61 Temperature measured from 4 pm to 6 pm at top point of the inside walls.

Assumptions for cost analysis:
AC: 1.5 tonnes, 5-star rated window AC;
Energy consumption of AC: 1.3 KW/hr;
Electricity cost: Rs. 6.75 per unit;
AC functioning for 10 hours a day with each month having 30 days;
Note:
Savings in lower temperatures (<24°C) – 6% for every 1°C; Savings in higher temperatures (>24°C) – 4% for every 1°C; No other electricity consumption in the household (The Energy and Resources Institute, TERI).

The energy saved by difference in temperature calculated as per the source: the energy and resources Institute (TERI) and the calculation are based on controlled conditions like steady ambient temperature. The total electricity bill saved is only an

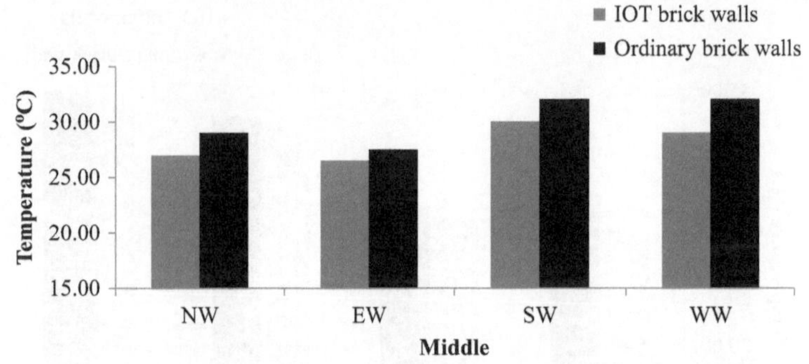

FIGURE 6.62 Temperature measured from 4 pm to 6 pm at middle point of the inside walls.

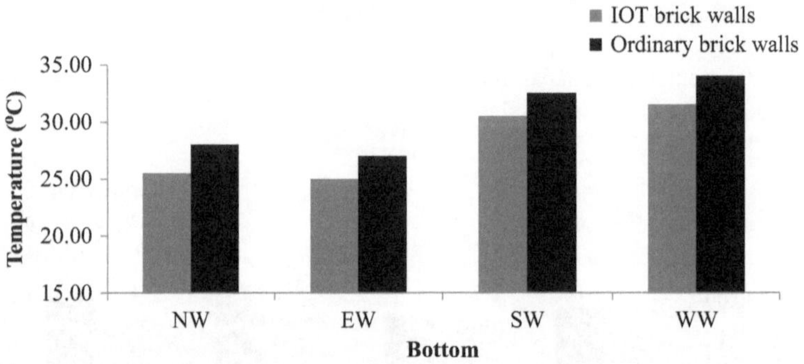

FIGURE 6.63 Temperature measured from 4 pm to 6 pm at bottom point of the inside walls.

TABLE 6.27
Temperature Measurement at 5 Minutes Interval After Plastering

Time	Ambient Temperature, °C	Ordinary brick room, °C	IOT–Perlite brick room, °C
0	33.80	33.50	31.20
5	33.40	33.80	31.40
10	33.50	33.80	31.50
15	33.80	33.60	31.50
20	33.60	33.50	31.50
25	33.50	33.50	31.30
30	33.20	33.60	31.20
35	33.50	33.70	31.20
40	33.80	33.80	31.30
45	33.60	33.60	31.20
50	33.80	33.50	31.00
55	33.50	33.40	30.90

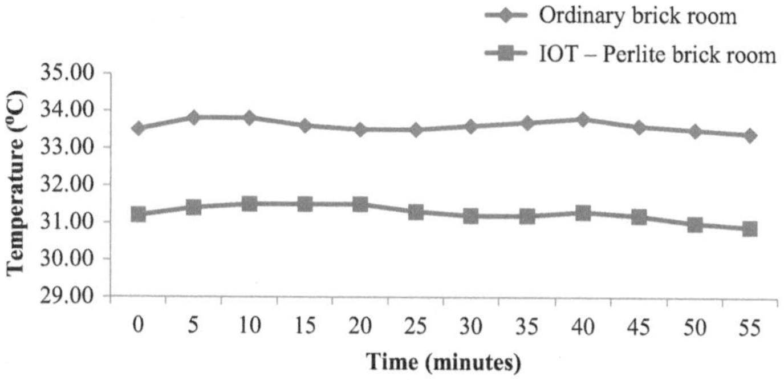

FIGURE 6.64 Temperature vs time after plastering of model rooms.

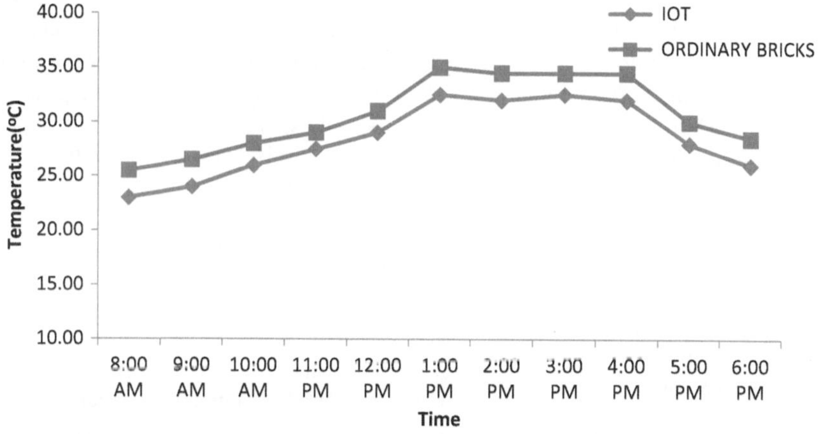

FIGURE 6.65 Temperature vs time for the day after plastering of model rooms.

TABLE 6.28
Energy and Cost Savings at Different AC Temperature Settings

Atmospheric temperature at 31°C

Room Temperature, °C	Total energy needed in Units, kWh/ day (when ambient temperature is 31°C) (a)	Electricity cost of using an AC (Rs. per day) (a) × 6.75 (b)	Electricity cost of using an AC (Rs. per month) (b) × 30 (c)
18	13.00	87.75	2,633
25	10.00	67.50	2,025
27	9.0	60.75	1,823

approximate amount and will vary depending on room insulation, actual room temperature, outside room temperature throughout the day, compressor run time, Energy Efficiency Ratio (EER), room size, total number of the people in the room, etc.; AC functioning for 8 months a year.

6.5 ECONOMIC FEASIBILITY STUDY

In order to assess the economic feasibility of thermal efficient bricks proposed in this study, a systematic study is carried out. All the materials used in the brick are converted to the units of marketable price. Table 6.29 shows the estimated cost details of brick with perlite and Table 6.30 shows the brick cost details without perlite. Based on the present price of the materials in the market, total cost of the brick is estimated.

Normal cost of the brick in the present market is Rs.14.00. Estimated cost of the brick with optimum combination is Rs. 13.54 and the cost of IOTs brick without perlite addition is Rs.9.35. In this study, IOT is brought from Bellary at distance of approximately 400 km for which the transportation cost is more and is considered with loading and unloading charges but if bricks are manufactured near the mining area then the considerable reduction will be achieved. In total there is marginal saving of 4% cost with perlite brick and a considerable savings of 38% with IOT bricks without perlite. The cost of IOTs and Perlite brick will be reduced, when the brick is prepared for construction in the vicinity of the mine area.

TABLE 6.29
Cost Analysis of Bricks (with Perlite)

Material Percentage	Cement (20%)	Sand (25%)	IOT (50%)	Perlite (5%)	Total (100)
Quantity of material required for each brick	0.67 Kg	0.85 Kg	1.68 Kg	0.16 Kg	–
Cost per kg, Rs.	6.00	0.75	0.70	30.00	–
Cost of material per brick, Rs.	4.02	0.64	1.18	4.80	10.64
Labour cost, Rs.	–	–	–	–	2.90
Total cost of the brick, Rs.	–	–	–	–	13.54

TABLE 6.30
Cost Analysis of Bricks (Without Perlite)

Material percentage	Cement (20%)	Sand (30%)	IOT (50%)	Total (100)
Quantity of material required for each brick	0.727 Kg	1.09 Kg	1.82 Kg	–
Cost per kg, Rs.	6.00	0.75	0.70	–
Cost of material per brick, Rs.	4.36	0.82	1.27	6.45
Labour cost, Rs.	–	–	–	2.90
Total cost of the brick, Rs.	–	–	–	9.35

6.6 DEVELOPMENT OF REGRESSION MODELS

Based on the large quantity of data generated, regression models were developed to predict various properties like density, compressive strength, water absorption and thermal conductivity. 70% of the data is used to develop models and the remaining data is used to validate the models.

6.6.1 Prediction of Density

In order to predict the density, the input parameters considered are IOT (I), sand (S), perlite (P) and cement (C) in terms of percentage, also the input parameters are varied so as to obtain the correlation between the input and output parameters. Using ANOVA analysis Equation-4.1 is developed and the regression analysis is performed between actual and predicted values using Equation-6.7 and the value of regression co-efficient (R^2) is found to be 0.83 which shows a very good correlation. From the analysis, it can be stated that cement is the most significant influence on density of brick followed by the other parameters IOT, sand and perlite. The parametric estimates are given in Table 6.31 and the analysis of variance (ANOVA) test summary given in Table 6.32 indicates that the model is robust as P values are less than

TABLE 6.31
Parametric Estimates for Density of IOT–Perlite Brick

Term	Coefficient	SE Coefficient	T-Value	P-Value	VIF
Constant	–27148	6831	–3.97	0.000	
IOT	288.9	68.2	4.24	0.000	2251.00
SAND	285.8	68.2	4.19	0.000	2626.00
PERLITE	201.3	68.6	2.93	0.006	76.00
CEMENT	303.3	68.3	4.44	0.000	301.00

TABLE 6.32
ANOVA Summary for Density of IOT–Perlite Brick

Source	DF	Adj SS	Adj MS	F-Value	P-Value
Regression	4	1460210	365053	39.26	0.000
IOT	1	166866	166866	17.94	0.000
SAND	1	163419	163419	17.57	0.000
PERLITE	1	80005	80005	8.60	0.006
CEMENT	1	183397	183397	19.72	0.000
Error	31	288277	9299		
Total	35	1748487			

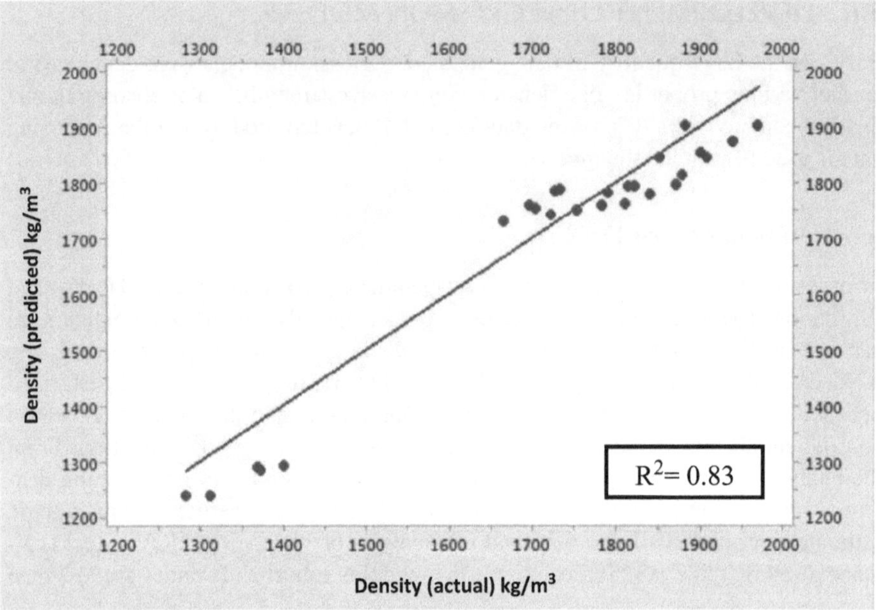

FIGURE 6.66 Relationship between predicted and measured values of density.

0.05. Figure 6.66 shows the regression fit between the observed values and predicted values.

$$\text{Density}\left(\text{kg/m}^3\right) = -27148 + 288.9\text{I} + 285.8\text{S} + 201.3\text{P} + 303.3\text{C}.... \quad (6.7)$$

where,

I = Iron ore tailings
S = Sand
P = Perlite
C = Cement

6.6.2 Prediction of Compressive Strength

Considering all the input parameters IOT (I), sand (S), perlite (P) and cement (C)percentage composition as whole, the output parameter compressive strength in terms of MPa is predicted as shown in Equation 6.8. R^2 value for compressive strength was found to be 0.88 between actual and predicted values. Results from regression analysis reveal that the parameter perlite has significant contribution and cement has least contribution on compressive strength of brick. The other two input parameters which influence the compressive strength are sand and IOT. Figure 6.67 shows the variation between observed and predicted values for the compressive strength (MPa). The parametric estimates are given in Table 6.33 and the analysis of variance (ANOVA)

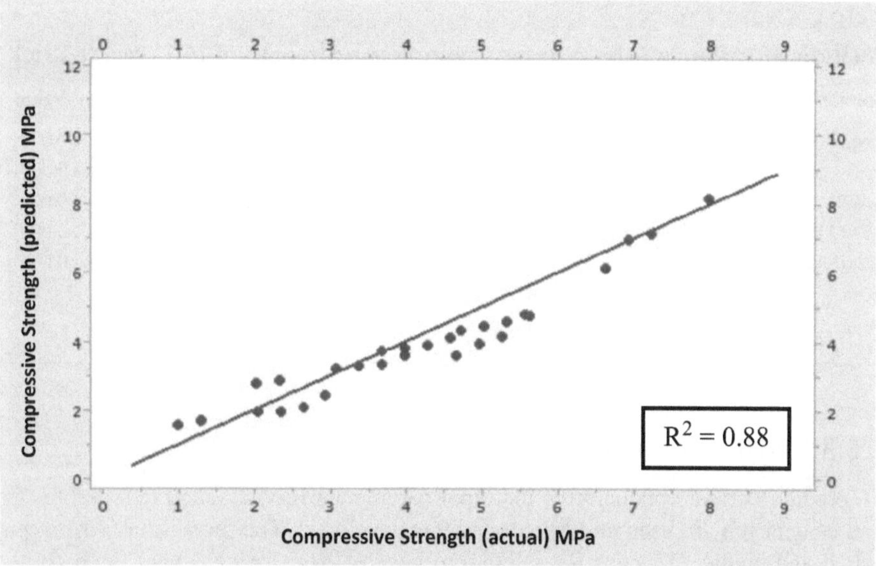

FIGURE 6.67 Relationship between predicted and measured values of compressive strength.

TABLE 6.33
Parametric Estimates for Compressive Strength of IOT–Perlite Brick

Term	Coefficient	SE Coefficient	T-Value	P-Value	VIF
Constant	220.8	63.5	3.48	0.002	
IOT	–2.173	0.634	–3.43	0.002	2251.00
SAND	–2.203	0.634	–3.48	0.002	2626.00
PERLITE	–3.062	0.638	–4.80	0.000	76.00
CEMENT	–1.873	0.635	–2.95	0.006	301.00

test summary given in Table 6.34 indicates that the model is robust. The regression equation is:

$$\text{Compressive Strength}(\text{MPa}) = 220.8 - 2.173\,\text{I} - 2.203\,\text{S} - 3.062\,\text{P} - 1.873\,\text{C}\ldots \quad (6.8)$$

where,

I = Iron ore tailings
S = Sand
P = Perlite
C = Cement

TABLE 6.34
Analysis of Variance (ANOVA) for Compressive Strength of IOT–Perlite Brick

Source	DF	Adj SS	Adj MS	F-Value	P-Value
Regression	4	189.973	47.4933	59.14	0.000
IOT	1	9.440	9.4404	11.76	0.002
SAND	1	9.709	9.7094	12.09	0.002
PERLITE	1	18.501	18.5009	23.04	0.000
CEMENT	1	6.990	6.9898	8.70	0.006
Error	31	24.895	0.8031		
Total	35	214.869			

6.6.3 Prediction of Thermal Conductivity

To predict thermal conductivity, the input parameters IOT (I), sand (S), perlite (P) and cement (C) are used and Equation 6.9 is developed. The regression fit between actual and predicted values R^2 is found to be 0.94, which shows very good correlation. Figure 6.68 shows the regression fit between the observed values and predicted values for output thermal conductivity. The parametric estimates are given in Table 6.35 and the analysis of variance (ANOVA) test summary given in Table 6.36

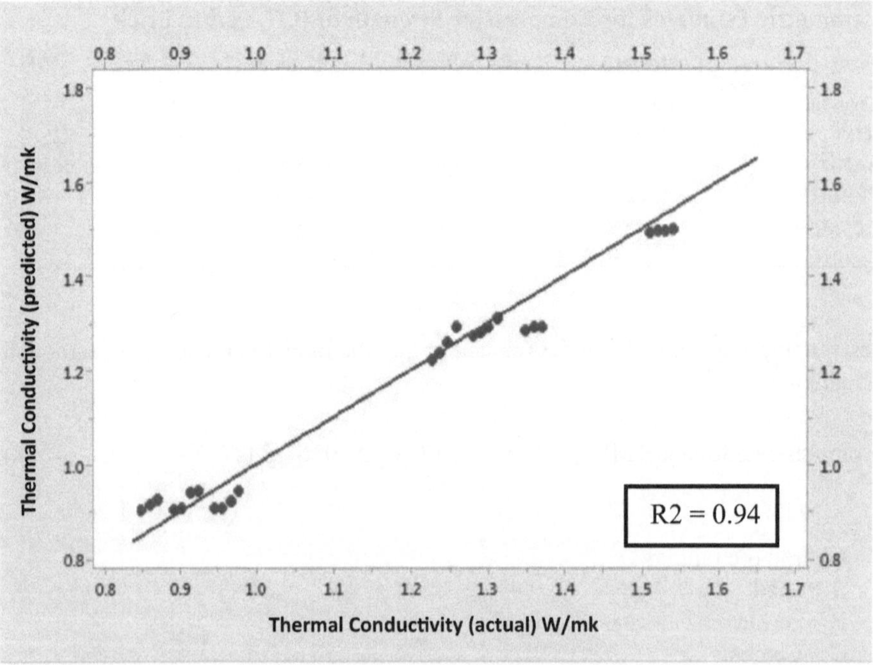

FIGURE 6.68 Relationship between predicted and measured values of thermal conductivity.

TABLE 6.35
Parametric Estimates for Thermal Conductivity of IOT–Perlite Brick

Term	Coefficient	SE Coefficient	T-Value	P-Value	VIF
Constant	-9.13	6.04	-1.51	0.141	
IOT	0.1061	0.0602	1.76	0.088	3241.05
SAND	0.1050	0.0602	1.75	0.091	3781.50
PERLITE	-0.0295	0.0604	-0.49	0.628	109.56
CEMENT	0.1158	0.0613	1.89	0.068	447.52

TABLE 6.36
ANOVA for Thermal Conductivity of IOT–Perlite Bricks

Source	DF	Adj SS	Adj MS	F-Value	P-Value
Regression	4	2.82423	0.706058	140.34	0.000
IOT	1	0.01563	0.015628	3.11	0.088
SAND	1	0.01532	0.015325	3.05	0.091
PERLITE	1	0.00120	0.001202	0.24	0.628
CEMENT	1	0.01797	0.017971	3.57	0.068
Error	31	0.15596	0.005031		
Total	35	2.98020			

shows that all the independent variables were significant at 94% confidence level. The regression equation is:

$$\text{Thermal Conductivity}\left(\text{W/mk}\right) = -9.13 + 0.1061\,\text{I} + 0.1050\,\text{S} - 0.0295\,\text{P} + 0.1158\,\text{C} \tag{6.9}$$

where,

I = Iron ore tailings
S = Sand
P = Perlite
C = Cement

6.7 SUMMARY

Physical properties of IOT are confirming to IS code specifications. Chemical composition of the IOT revealed that, it consists of major elements such as high percentage of silicon dioxide (SiO_2), Aluminium oxide (Al_2O_3), Iron oxide (Fe_2O_3) and Calcium oxide (CaO). So, density of bricks increased with an increase in percentage of IOT. The density of brick can be reduced by the addition of perlite which is proved in the test results. Compressive strength of bricks increased with increase in percentage of IOT and cement. No perceptible deposit of efflorescence was observed when there is 5% of perlite in brick. In the absence of perlite in brick, efflorescence was present. So, perlite addition is an added advantage in this regard. Thermal conductivity of the brick is substantially reduced by adding perlite.

Brick of IOT and perlite have achieved good results in quality assessment tests like UCS, thermal conductivity and in durability tests such as water absorption, density and efflorescence. The mixture of 5% perlite with 50% IOT, 20% cement and 25% sand has given an optimum proportion for brick production to get better results on all tests.

From pilot-scale studies, it was found that the heat from the atmosphere is arrested at outside surface of the IOT–perlite brick wall better than ordinary brick wall before passing through the bricks. It showed lower temperature at outside surface of the walls than the ordinary brick walls. It is noticed that inside temperature of the room, constructed by IOT–perlite bricks on an average is less by 2°C compared to ordinary bricks. After the plastering of the wall, IOT–perlite brick room had lower room temperature by 2.5°C compared to ordinary brick room, further reduction in room temperature was observed when it is compared to without plastering the IOT–perlite brick wall, which results around 8% of saving of electricity.

It was found that the manufacturing cost of IOT–perlite brick will be approximately 4% less than the cost of the bricks available in the market. But the bricks without perlite that can be used in the load bearing walls will cost 38% less than the ordinary bricks which is substantial in reducing the cost of construction. Though the economic advantage is nominal in IOT–perlite bricks, the use of IOT in bricks is a great advantage for construction as well as for mining sector. In addition to the above, usage of mine waste in the form of IOT reduces the burden on depleting naturally available river sand and clay, on the other hand, disposal and maintenance cost of tailings will be a big saving to the mining and mineral industry and immediate use of IOT without storing reduces the environmental effect on the surroundings. As the bricks are non-fired, it results in lower embodied energy and also acts as energy conservative. Manufacturing of brick will not have any effect on the environment.

REFERENCES

Arici, M., Yilma, B. and Karabay, H. 2016. Investigation of heat insulation performance of hollow clay bricks filled with perlite. *International Conference on Computational and Experimental Science and Engineering (ICCESEN 2015)*, 130, 266–268.

ArunRaja, L., Arumugam, M. and Malaichamy, K. 2017. Mechanical properties of light weight bricks using perlite and lime material. *SSRG International Journal of Civil Engineering*, special issue, 106–113.

Ayudhya, B.I.N. 2011. Compressive and splitting tensile strength of autoclaved aerated concrete (ACC) containing perlite aggregate and polypropylene fiber subjected to high temperatures. *Songklanakar in Journal of Science and Technology*, 33(5), 555–563.

Benk, A. and Coban, A. 2012. Possibility of production lightweight heat insulating bricks from pumice and H_3PO_4 or NH_4NO_3 hardened molasses binder. *Ceramics International*, 38(3), 2283–2293.

Bindiganavile, V., Batool, F. and Suresh, N. 2012. Effect of flyash on thermal properties of cement based foams evaluated by transient plane heat source. *The Indian Concrete Journal*, 86(11), 7–14.

Bullibabu, K. and Ali, A.M.D. 2018. Production and characterization of low thermal conductivity clay bricks by admixture of bagasse and perlite. *International Journal of Mechanical and Production Engineering Research and Development (IJMPERD)*, 8(3), 809–816.

Bulut, U. 2010. Use of perlite as a pozzolanic addition in lime mortar. *Gazi University Journal Science*, 23(3), 305–313.

Bureau of Indian Standards (BIS). Method for the determination of thermal conductivity of thermal insulation materials – Two slab, guarded hot plate method. IS: 3346: New Delhi, India; 1980a.

Bureau of Indian Standards (BIS). Sand for masonry mortars specification. IS: 2116: New Delhi, India; 1980b.

Bureau of Indian Standards (BIS). Determination of compressive strength – Method of tests of burnt clay building bricks. IS: 3495 (Part 1): New Delhi, India; 1992a.

Bureau of Indian Standards (BIS). Determination of water absorption – Method of tests of burnt clay building bricks. IS: 3495 (Part 2): New Delhi, India; 1992b.

Bureau of Indian Standards (BIS). Determination of efflorescence – Method of tests of burnt clay building bricks. IS: 3495 (Part 3): New Delhi, India; 1992c.

Bureau of Indian Standards (BIS). 53 grade ordinary portland cement specification. IS: 12269: New Delhi, India; 2013.

Celik, A.G. (2014). Investigation on characteristic properties of potassium borate and sodium borate blended perlite bricks. *Journal of Cleaner Production*, 102, 88–95.

Celik, M. H. and Durmaz, M. 2012. The study of compressive strength in different cement types and dosages of concretes made by using 60% pumice and 40% perlite. *Contemporary Engineering Sciences*, 5(6), 287–293.

Chaouki, S., Abderrahman, A., Iz-Eddine, E.I. and Amrani. 2013. Production of porous fire brick from mixture of clay and recycled refractory waste with expanded perlite addition. *Journal of Material Environmental Science*, 4(6), 981–986.

Demir, M. and Orhan. 2004. An investigation on the production of construction brick with perlite addition. *Key Engineering Materials*, 264–268, 2161–2164.

Figen, B., Aynur, U and Yucel, H.L. 2010. Development of the insulation materials from coal, flyash, perlite, clay and linseed oil. *Ceramics – Silikaty*, 54(2), 182–191.

Georgiev, Angel, Albena, Yoleva, Stoyan, Djambazov, Dimitar, D. and Ivanova, D.V. 2017. Effect of expanded vermiculite and expanded perlite as pore forming additives on the physical properties and thermal conductivity of porous clay bricks. *Journal of Chemical Technology and Metallurgy*, 53(2), 275–280.

Gonzalez, E., Galan, A., Miras, M.A. and Vazquez 2011. CO_2 emissions derived from raw materials used in brick factories: Applications to andalusia (Southern Spain). *Applied by Clay Science*, 52(3), 193–198.

Jedidi, M., Benjeddou, O. and Soussi, Ch. 2015. Effect of expanded perlite aggregate dosage on properties of lightweight concrete. *Jordan Journal of Civil Engineering*, 9(3), 278–291.

Lanzon, M. and Garcia-Ruiz, P.A. 2008. Light-weight cement mortars: Advantages and inconveniences of expanded perlite and its influence on fresh and hardened state and durability. *Construction and Building Materials*, 22(8), 1798–1806.

Li, Runfeng, Zhou, Y., Cuiwei, Li, Li, Shibo S. and Zhenying, H. 2019. Recycling of industrial waste iron ore tailings in porous bricks with low thermal conductivity. *Construction and Building Materials*, 213, 43–50.

Michael, C., Steven, H.M. and Rapport, J.. 2009. The building bricks of sustainability. *The Construction Specifier*, 1, 30–40. https://www.researchgate.net/publication/285295384_The_building_brick_of_sustainability

Mohajer, A.S., Ghanbarnezhad, S., Sharifi, L., Mirhosseini, S.H. and Irvani, Y.M. 2014. Production of lightweight refractory insulation panels on the basis of perlite. *International Journal of Research Science and Management*, 1(1), 1–12.

Naveen, J., Bullibabu, K., Veeranjaneyulu, K., and Abidali. 2018. Production and analysis of composite construction materials with admixture of coal – bagasse based fly ash and perlite by ANSYS approach. *International Research Journal of Engineering & Technology*, 5(8), 735–739.

Samar, M and Saxena, S. 2016. Study of chemical and physical properties of perlite and its application in India. *International Journal of Science Technology and Management*, 5(4), 70–80.

Satakhun, D., Tanakorn, P., Vanchai, S., Chudapak, D. and Prinya, Ch. 2018. Portland cement containing fly ash, expanded perlite, and plasticizer for masonry and plastering mortars. *International Journal of GEOMATE*, 15(48), 107–113.

Sengul, O., Zaizi, S., Karaosmanogu, F. and Tasdemir, M. A. 2011. Effect of expanded perlite on the mechanical properties and thermal conductivity of lightweight Concrete. *Energy and Buildings*, 43(2–3), 671–676.

Shankarananth, S. and Jaivignesh, B. 2016. Experimental study on the use of glass powder, GGBS, & perlite in fly ash brick. *International Journal of Advanced Research*, 4(4), 1381–1387.

Toledo, R., DosSantos, D.R., Faria, R.T., Carrio, J.G., Avler, L.T. and Vargas, H. 2004. Gas release during clay firing and evolution of ceramic properties. *Applied Clay Science*, 27, 151–157.

Topcu, Iiker Beki and Isikdag, Burak. 2006. Manufacture of high heat conductivity resistant clay bricks containing perlite. *Building and Environment*, 42(10), 3540–3546.

Uluer, O., Aktas, M., Karaagac, I., Durmus, G., Khanlari, A., Agbulut, U. and Celik, D.N. 2018. Mathematical calculation and experimental investigation of expanded perlite based heat insulation materials thermal conductivity values. *Journal of Thermal Engineering*, 4(5), 2274–2286.

Xu, F., Peng, C., Zhu, J. and Chen, J. 2016. Design and evaluation of polyester fiber and SBR latex compound – Modified perlite mortar with rubber powder. *Construction and Building Materials*, 127, 751–761.

Zulkifeli, M. and Saman, H. 2016. Compressive and flexural strength of expanded perlite aggregate mortar subjected to high temperatures. *International Conference on Applied Physics and Engineering (ICAPE2016)*, Malaysia.

WEBSITES

http://www.keltechenergies.com/perlite-products.html, Accessed on 5th Dec. 2018.

http://www.hertig.com.ar/PDFWeb/Productos/TemperaturaAireHumedadPh/Termometros/VA6502.pdf, Accessed on 18th Aug 2019.

https://www.primusthai.com/uploads/files_th/CENTER350_352.pdf, Accessed on 18th Aug 2019.

https://www.acchelp.in/purchase_managers_certificate.html, Accessed on 14th April 2020.

https://www.researchgate.net/profile/Isaac_Animasaun/post/How_can_i_calculate_thermal_conductivity_of_porous_ceramics_without_effect_of_porosity/attachment/5cefc672cfe4a7968da5529c/AS%3A764228906541056%401559217778018/download/chapter-16.pdf, Accessed on 10th Sep 2020.

BIBLIOGRAPHY

Abdulrahman, H. S. 2015. Potential use of iron ore tailings in sand Crete block making. *International Journal of Research in Engineering and Technology*, 4(4), 409–414.

Anand, K. and Singh, O. 2016. Indian iron ore scenario: low grade iron ore beneficiation. MECON Limited, Ranchi, http://www.meconlimited.co.in/writereaddata/MIST_2016/sesn/tech_1/5.pdf, Accessed September 2019.

Beatryz, C., Leonardo, M., Pedroto, G., Mauricio, P.F., Jose, F., Carlos, L., Carlos, R.M.F. Vieira. Anderson, A. Pacheco. Afonso, R.G. Azevedo de. 2019. Technical and environmental assessment of the incorporation of iron ore tailings in construction clay bricks. *Construction and Building Materials*, 227.

Carrasco, E.V.M., Mantilla, J.N.R., Esposito, T. and Moreira, L.E. 2013. Compression performance of walls of interlocking bricks made of iron ore byproducts and cement. *International Journal of Civil & Environmental Engineering*, 13(3), 56–62.

Central Pollution Control Board, 2007. Compressive industrial document on iron ore mining. *Ministry of Environment and Forests, Govt. of India, New Delhi*, http://www.indiaenvironmentportal.org.in/files/iron_ore_mining.pdf.

Chuanmeng, Y., Chong, C. and Juan, Q. (2014). "Recycling of low-silicon iron ore tailings in the production of lightweight aggregates." *Ceramics International*, 41(1), 1213–1221.

Das, S.K., Kumar, S. and Rao, R.P. 2000. Exploitation of iron ore tailing for the development of ceramic tiles. *Waste Management*, 20(8), 725–729.

Ellen, Bau. 2010. Black carbon from brick kilns. *Environmental Science*, Presentation for Clean Air Task Force.

Francis, A., Kuranchie, S.K., Shukla, D. and Habibi. 2014. Utilization of iron ore tailings for the production of geo polymer bricks. *International Journal of Mining, Reclamation and Environment*, 30(2), 92–114.

Giri, B.S.V. and Krishnaiah, S. 2018. Manufacturing of eco-friendly brick: A critical review. *International Journal of Computational Engineering Research (IJCER)*, 8, 24–32.

Government of India, Ministry of Mines India Bureau of Mines (2018). *Indian Minerals Yearbook 2017, Perlite* (Part-III: Mineral Reviews), 56th Edition.

Government of India, Ministry of Mines India Bureau of Mines (2019). *Indian Minerals Yearbook 2018, Iron Ore* (Part-III: Mineral Reviews), 57th Edition.

Hammond, A.A. 1988. Mining and quarrying wastes: A critical review. *Engineering Geology*, 25 (1), 17–31.

Kshitija, K., Vikrant, S., Mujahed, S., Mahesh, K., Mahendra, C. and Sagar, S. 2015. Use of iron ore tailings as a construction material. *International Journal of Conceptions on Mechanical and Civil Engineering*, 1(2), 1–21.

Kumar, S., Kumar, R. and Amitava, B. 2006. Innovative methodologies for the utilization waste from metallurgical and allied industries. *Resources Conservation and Recycling*, 48(4), 301–314.

Likith, N.P., Manjunatha, Siddesh, S.S., Manjunath, K.M., Hadimani, S. and Shivkumar, B. 2017. Manufacturing of building blocks by utilizing of iron ore tailings. *International Journal of Engineering Science and Computing*, 7(5), 12274–12277.

Mohanty, M. Dhal, N.K. Patra, P. Das, B. and Reddy, P.S.R. 2010. Phytoremediation: A novel approach for utilization of iron-ore wastes. *In Reviews of Environmental Contamination and Toxicology*, 206, 29–47.

Nagaraj, H.B., Rajesh, A. and Sravan, M.V. 2016. Influence of soil gradation proportion and combination of admixtures on the properties and durability of CSEBS. *Construction and Building Materials*, 110, 135–144.

Namratha, M. Aruna, R. and Saraswathi, M. 2014. Strengthening of fly ash bricks by ironite. *IOSR Journal of Mechanical and Civil Engineering*, 11(3), 21–26.

Nithiya, R. Chris. Anto, L. Vinodh, K.R. and Anbalagan, C. 2016. Experimental investigation on bricks by using various waste materials. *International journals of latest trends in engineering and technology (IJLTET)*, 6(3), 395–402.

Prahallada, M.C. and Shanmuka, K.N. 2014. Stabilized iron ore tailings block an – environmental friendly construction material. *International Journal of IT, Engineering and Applied Sciences Research*, 3(4), 2319–4413.

Prem, K.W. P., Ananthya, M.B. and Vijay, K. 2014. Effect of replacing sand by iron ore tailings on the compressive strength of concrete and flexural strength of reinforced concrete blocks. *International Journal on Research in Engineering and Technology*, 3(7), 1374–1376.

Ravi, K.C.M., Kumar, A., Prashanth, M.H. and Reddy, V.D. 2012. Experimental studies on iron ore tailings based interlocking paves blocks. *International Journal of Earth Sciences and Engineering*, 5(3), 501–504.

Skanda, K.B.N. 2014. Utilzation of iron ore tailings as replacement to fine aggregates in cement concrete pavements. *International Journal of Research in Engineering and Technology*, 3(7), 369–376.

Sujing, Z., Junjiang, F. and Wei, S. 2013. Utilization of iron ore tailings as fine aggregate in ultra – high performance concrete. *Construction and Building Materials*, 50, 540–548.

Ugama, T. I. and Ejeh, S. P. 2014. Iron ore tailings as fine aggregate in mortar used for Masonry. *International Journal of Advances in Engineering and Technology*, 7(4), 1170–1178.

Ullas, S.N., Reddy, V.B.V. and Rao, N.K.S. 2010. Characteristics of masonry units from iron ore tailings. *International Conference on Sustainable Built Environment, Kandy*, 108–114.

Xiaoyan, H., Ravi, R. and Victor, Li. 2013. Feasibility study of developing green ECC using iron ore tailings powder as cement replacement. *Journal of Materials in Civil Engineering*, 25(7), 923–931.

Yisa, G. L., Akanbi, D. O., Agbonkhese, O., Ahmed, M. K. and Sani, J. E. 2016. Effect of iron ore tailing on compressive strength of manufactured laterite bricks and its reliability estimate. *Civil and Environmental Research*, 8(8), 49–58. https://shodhganga.inflibnet.ac.in/bitstream/10603/152177/16/9.%20chapter-1.pdf, Accessed on 20 September 2017.

Yongliang, C., Yimin, Z., Tiejun, C., Yunliang, Z. and Shenxu, B. 2010. Preparation of eco-friendly construction bricks from hematite tailings. *Construction and Building Materials*, 25, 2107–2111.

7 Iron Ore Mine Waste and Tailings as Aggregates in Concrete

Gayana B. C. and Ram Chandar Karra

CONTENTS

7.1 INTRODUCTION

The increasing trend of solid waste generation has given scope for its ecological disposal due to its hazardous social and environmental concerns. One of the methods to minimize these issues is recycling of the industry wastes. Construction industry has

DOI: 10.1201/9781003268499-7

a great prospective to absorb various industrial residues provided the recycled materials are properly characterized before its applications as a construction material. Iron is the world's commonly used metal – steel, where iron ore is the raw material, representing almost 95% of all metals used annually.

The scope of research is towards green concrete and efforts are being made by the researches to find the suitability of marginal materials to replace cement and natural aggregates in concrete. Generally, the marginal materials used for replacement of cement are Granulated Blast Furnace Slag (GGBS), fly ash, silica, metakaoline, alccofine, etc.

Need for aggregate replacement is demanding for the present scenario of construction industry as the mining of sand is a depleting resource. The rapid development of infrastructure has increased the need of sand mining due to which the river bed is over exploited and the need to recycle materials is in demand. Iron ore tailings (IOT) is the processed waste during the steel processing and is disposed in tailing ponds. A few studies by some researchers suggested that IOT can be considered as a construction material. The various environmental issues of sand mining are depletion of virgin deposits, water table lowering, collapsing of river banks and water pollution (Kang et al. 2011). An increase in the production of iron ore for economic development worldwide has been generating massive amount of IOT, which are frequently being discarded as wastes. This has led to serious environmental deterioration. A statistical survey has shown that 5 to 7 billion tonnes of IOT were produced yearly worldwide (Edraki et al. 2014). In spite of such huge amount of IOT stockpiled as waste, its safe disposal or utilization has remained a major unsolved and challenging task for iron ore industries (Yu et al. 2011).There are various other mine wastes which need to be explored as marginal materials in construction industry. Ram Chandar et al. (Ram Chandar et al. 2016a, 2016b) investigated the use of laterite and sandstone as partial replacement of sand, it was observed that sandstone enhanced the strength properties of concrete and laterite did not show much improvement on strength properties of concrete. Filho et al. (2017) evaluated the use of IOTs as fine aggregates in interlock concrete blocks. The physical and mechanical properties were superior to the conventional interlock paver blocks. Castro Bastos et al. (2016) evaluated the feasibility of IOT as a road material. The IOT was chemically stabilized using cement, lime, steel-making slag as binder consisting of 1%, 2%, 5% and 10% binder content, respectively. XRF, XRD, leaching, compaction and CBR tests were conducted and the results suggested that IOT with chemical stabilization is feasible to be considered as a road paving material. Mohammad and Bareither (2017) evaluated the binder amendment of Unconfined Compressive Strength (UCS) of mine tailings in the application of earthworks. A review on the use of iron ore mine waste and tailings and other industrial waste materials as replacement for aggregates with various admixtures to enhance the properties of concrete was done (Gayana and Ram Chandar 2018).

Various researchers have experimented the use of IOT as replacement of fine aggregates in concrete by partially replacing cement with fly ash, GGBS, silica fumes, pond ash, alccofine etc. The present study aims to enhance the properties of IOT concrete with addition of alccofine as partial replacement of cement. Based on the literature review discussed, alccofine of 10% is considered as cement replacement.

The optimization of the concrete mixes was observed for varying IOT replacement (0%, 10%, 20%, 30%, 40% and 50%) as fine aggregates for different curing days (3, 7, 28 and 56 days) and water–cement ratios (0.35, 0.40 and 0.45).

7.2 MATERIALS AND METHODS

7.2.1 Cement and Virgin Aggregates

Ordinary Portland Cement (OPC) of Grade 53 with specific gravity 3.14 and initial and final setting times were 60 and 450 minutes, respectively. The properties of cement are listed in Table 7.1 and compared with the IS 12269:2013 and are found to be within the limits.

The coarse and fine aggregates used are natural and locally available materials in south western part of India. Table 7.2 shows the physical properties of the aggregates as per IS 383-1970.

TABLE 7.1
Physical Properties of OPC and Alccofine

Sl No	Properties	OPC-53	Requirement as per IS 12269:2013	Remarks
1	Specific gravity	3.14	–	The cement satisfies the requirement for 53 grade OPC stipulated by IS 12269:2013. Tests are conducted as per guidelines of IS 4031.
2	Standard consistency (%)	29	–	
3	Fineness (m^2/kg)	300	Should not be less than 225 m^2/kg	
4	Initial setting time (min)	60	Should not be less than 30 minutes	
5	Final setting time (min)	450	Should not be more than 600 minutes	
6	Soundness (mm) (By Le Chatelier Mould)	2	Should not exceed 10 mm	

TABLE 7.2
Physical Properties of Natural and Marginal Aggregates

Sl. No	Property	Coarse Aggregates	Fine Aggregates	Iron Ore Waste Rock (WR)	Iron Ore Tailings (IOT)
1	Specific gravity	2.8	2.7	2.84	3.31
2	Bulk density-loose (kg/m^3)	1370	1440	1354	1684
3	Bulk density-compacted (kg/m^3)	1670	1770	1658	1816
4	Moisture content	Nil	Nil	NIL	3.9
5	Water absorption (%)	0.50	0.10	0.45	2.29

7.2.2 Water and Superplasticizer

The amount of water in a concrete mix has a direct effect on the strength development of the mixture. Water must be added as per the mix design to lubricate the solids in the mixture. Tap water was used for mixing concrete. Sulphonated naphthalene formaldehyde polymer admixture (Conplast SP 430) is used in the present study to improve the workability of concrete. The properties of Conplast SP 430 are as follows: its specific gravity is 1.20, chloride content is nil, solid content is 40%, 10–40°C is the operating temperature and the colour of admixture is dark brown liquid.

7.2.3 Iron Ore Waste Rock

Iron ore waste rock (WR) was collected from an iron ore mine located in southern part of India). Sampling was done as per IS 2386 (Part III, IV): 1963. The physical properties of the WR are given in Table 7.2. The WR is considered as replacement of coarse aggregates in concrete.

7.2.4 Iron Ore Tailings

IOT is obtained from tailings pond of an iron ore mine located in Southern part of India. The samples were collected by random sampling at various locations from the tailing pond as per IS 1199-1959. The basic physical properties were determined to assess the suitability of IOT as aggregate in concrete which are given in Chapter 6.

7.2.5 Chemical Composition of WR and IOT using X-Ray Florescence

The chemical composition of WR and IOT is analysed using XRF and the details are shown in Table 7.3. It is observed that, the materials consists of major elements viz., high percentage of Silicon dioxide (SiO_2), Iron oxide (Fe_2O_3), Aluminium oxide (Al_2O_3), Calcium oxide (CaO).

7.2.6 Leaching

The heavy metal concentration in IOT was As, Ba, Cd, Cr, Pb, Se, Ag, Zn and Cu, which confirms that the materials were non-hazardous. Toxicity Characteristic Leaching Procedure (TCLP), established by United States Environmental Protection Agency (US EPA), was used in estimating the heavy metal concentration from IOT material. The experimental procedure as per US-EPA guidelines was followed and the IOT materials were mixed with deionized water at a Liquid-Solid ratio (L/S) of 20:1 and 30 rpm agitation for 24 hours. Later leachates were extracted and filtered, thereby determining the heavy metal concentration by Inductively Coupled Plasma-Mass Spectrometry (ICP-MS). The heavy metal concentration in IOT determined is tabulated in Table 7.4 and were compared to the regulatory limits. Based on the results, the IOT material used in the present study can be considered as non-hazardous mine waste and can be utilized as a construction material.

TABLE 7.3
Chemical Composition by XRF of WR and IOT

Chemical Composition (%)	WR	IOT
SiO_2	51.861	49.750
Al_2O_3	4.643	7.377
Fe_2O_3	23.52	27.030
CaO	4.643	4.057
MnO	0.803	0.171
K_2O	0.131	0.500
ZnO	–	0.100
CuO	0.141	0.200
PbO	–	0.400
MgO	6.347	3.109
Na_2O	1.402	–
P_2O_5	0.179	–
SO_3	0.39	–
TiO_2	1.449	–
Cr_2O_3	0.104	–
NiO	0.105	–
pH	7.5	8.030
Electrical conductivity (mS)	0.238	0.329

TABLE 7.4
Heavy Metals Concentration in IOT

Elements	Regulatory Limits (mg/l)	Concentration in IOT (mg/l) (Present Study)	Concentration in IOT (mg/l) (Ali., et al., 2016)
As	0.05	0.005	0.002
Ba	1.00	0.0006	0.0008
Cd	0.01	0.004	0.002
Cr	0.05	0.0003	0.0002
Pb	0.05	0.0010	0.0008
Se	0.01	0.002	0.001
Ag	0.05	0.01	0.01
Zn	0.50	0.0009	0.0005

7.3 MIX DESIGN

The nominal mix ratio was designed for the marginal materials as replacement for coarse and fine aggregates as discussed in this section:

Designing an appropriate mix for a particular strength and workability is very important for assessing the properties of the materials used in concrete. Each material will influence the properties of concrete in its own way. Concrete mix for M40 grade was designed following the IS 10262:2009. The total binder content is restricted to 425 kg/m^3, with water/binder ratio of 0.35, 0.40 and 0.45 for coarse aggregate: fine aggregate ratio of 0.64:0.36. The mixes are designed to achieve a slump value of 25–50 mm. A super plasticizer quantity of 1.0% (by weight of binder content) is added to the mix to arrive at the designated slump. Mix proportions for concrete mixes are estimated for three different compositions.

WR and IOT were replaced partially as coarse and fine aggregates, respectively, in two different mixes, i.e., 0%, 10%, 20%, 30%, 40% and 50% for 3, 7 and 28 days. This set of mixes is varied with varying w/c from 0.35, 0.40 and 0.45 as shown in Tables 7.5 and 7.6.

Six batches of mix proportions were prepared for each type of w/c; therefore, a total of 18 batches were prepared for three w/c ratios considered for varying mix proportions in the present study (Table 7.5). The different mixes with WR are 0%, 10%, 20%, 30%, 40% and 50%.

Six batches of mix proportions were prepared for each type of w/c; therefore, a total of 18 batches were prepared for three w/c ratios considered for varying mix proportions in the present study (Table 7.6). The different mixes with IOT are 0%, 10%, 20%, 30%, 40% and 50%.

TABLE 7.5
Mix Proportion for Waste Rock (WR) – Concrete

Mix	Cement (Kg)	Coarse Aggregates (Kg)	Fine Aggregates (Kg)	Water–Cement Ratio	Superplasticizer (% by wt. of Cement)	WR (Kg)
WR1-0	445	1066.00	789	0.35	0.5	–
WR1-10		946.25			0.5	183.50
WR1-20		858.32			0.5	214.58
WR1-30		733.05			0.5	328.30
WR1-40		647.91			0.5	431.94
WR1-50		541.67			0.5	541.67
WR2-0	445	1081.00	834	0.40	0.5	–
WR2-10		954.21			0.5	196.85
WR2-20		864.86			0.5	223.30
WR2-30		756.76			0.5	334.90
WR2-40		648.65			0.5	446.60
WR2-50		540.54			0.5	558.20
WR3-0	445	1103.00	85061	0.45	0.5	–
WR3-10		985.52			0.5	201.20
WR3-20		887.92			0.5	221.98
WR3-30		779.45			0.5	334.04
WR3-40		670.26			0.5	446.84
WR3-50		560.34			0.5	560.34

TABLE 7.6
Mix Proportion for Iron Ore Tailings (IOT) – Concrete

Mix	Cement (Kg)	Alccofine (Kg)	Coarse Aggregates (Kg)	Fine Aggregates (Kg)	Water–Cement Ratio	Super-Plasticizer (% by wt. of Cement)	IOT (Kg)
IOT1-0	445	50	1066	789.00	0.35	0.5	–
IOT1-10				711.00		1	97.00
IOT1-20				660.00		1	165.00
IOT1-30				590.10		1	252.90
IOT1-40				516.60		1	344.40
IOT1-50				439.50		1	439.50
IOT2-0	386	43	1120	796.00	0.40	0.5	–
IOT2-10				716.00		1	97.50
IOT2-20				636.85		1	195.18
IOT2-30				557.25		1	292.76
IOT2-40				477.64		1	390.36
IOT2-50				398.03		1	487.96
IOT3-0	346	39	1116.5	861.30	0.45	0.5	-
IOT3-10				775.00		1	105.58
IOT3-20				720.16		1	180.04
IOT3-30				641.69		1	275.01
IOT3-40				563.48		1	375.65
IOT3-50				479.30		1	479.30

7.4 EXPERIMENTAL INVESTIGATIONS

7.4.1 Specimen Preparation and Curing Conditions

The test blocks were prepared with concrete samples casted in different moulds depending upon the test requirements. Cubes of dimension 100 mm × 100 mm × 100 mm accounting to 162 cubes per mix proportion were casted for compressive strength (CS) each for WR concrete and IOT concrete. 54 cylinders of 150 mm diameter and 300 mm length were casted for splitting tensile strength (STS) and 54 beams of dimension 500 mm × 100 mm × 100 mm were prepared for testing flexural strength (FS) for varying w/c each for WR concrete and IOT concrete. Fresh concrete was used for the slump tests. It should be noted that among the 162 cubes prepared for compression tests, 3 replicate cubes each were cured for 4 different curing days (3, 7 and 28 days) prior to testing compressive strength and 3 cylinders and 3 beams for each mix proportion were considered at 28 days of curing prior to testing splitting tensile strength and flexural strength for 3 different w/c for WR and IOT concrete mixes.

The concrete mix was carried out using a drum mixer of 150 kg capacity. The mixer was hand loaded and the duration of mixing was about 2.5 to 3.0 minutes after addition of all the materials, i.e., cement, sand, coarse aggregates, WR, superplasticizer and water as per the mix design, until a homogenous concrete mix is attained for WR mixes and cement, sand, coarse aggregates, IOT, superplasticizer and water as per the mix design, until a homogenous concrete mix is attained for IOT mixes. The procedure for mixing adhered as per IS 516-1959. After the mixing operation, the materials were immediately transferred to the tray and the workability of fresh concrete was determined. The concrete was placed in slump cone in three layers and each layer was given 25 strokes using tamping rod to compact the concrete and to reduce the air voids (IS 516-1959).

7.4.2 Workability of Concrete

The fresh and hardened state of IOT-alccofine concrete mixes is assessed. The workability of concrete was characterized using the slump cone test. Later, the fresh concrete was placed in desired moulds and dried for 24 hours and then demoulded and water cured for the desired curing ages. The samples are weighed and the dimensions are recorded to determine the density as per IS 1199 (BIS 1959).

7.4.2.1 Effect of WR on Workability of Concrete

The target slump was attained as per the mix design for M40 grade for concrete with WR as coarse aggregates, i.e., 25–50 mm. The results are depicted in Table 7.7. Workability of concrete decreased with increasing percentage of WR for 0.35 w/c. Slump decreased with reference to the control mix by 8.57%, 14.29%, 20.00%, 20.00%, 28.57% for 10%, 20%, 30%, 40% and 50%, WR replaced as coarse aggregates, respectively. Slump values decreased with reference to the control concrete by 7.89%, 13.16%, 21.05%, 21.05% and 26.32% with WR replaced from 10 to 50% for 0.40 w/c. Similarly, for 0.45 w/c, slump decreased with reference to the control mix by 7.50%, 12.50%, 20.00%, 20.00% and 25.00%. WR replaced by 10% did not show much significance but with increase in WR percentage, significant increase in workability was observed. In comparison within the w/c, slump increased with increase in w/c from 0.35 to 0.45. Maximum slump of 50 mm was attained for control mix at 0.45 w/c compared to 0.35 and 0.40 w/c.

TABLE 7.7
Slump Values of Waste Rock (WR) Concrete at Varying w/c

Mix Proportion	Slump (mm)		
w/c	0.35	0.4	0.45
WR0-0	35	38	40
WR0-10	32	35	37
WR0-20	30	33	35
WR0-30	28	30	32
WR0-40	28	30	32
WR0-50	25	28	30

TABLE 7.8
Slump Values of Iron Ore Tailings (IOT) Concrete at Varying w/c

	w/c		
	0.35	**0.4**	**0.45**
Mix Percentage		**Slump (mm)**	
IOT-0	30	33	35
IOT-10	28	30	33
IOT-20	26	29	31
IOT-30	22	25	28
IOT-40	15	20	25
IOT-50	0	0	10

7.4.2.2 Effect of IOT on Workability of Concrete

With reference to the control mix, decrease in workability was observed with increase in IOT from IOT3-0 to IOT3-50 as illustrated in Table 7.8. Slump values were maximum for control mix concrete and gradually decreased with increase in IOT percentage and very low workability was observed with 50% replacement of IOT. With reference to results within w/c, an increase in workability was observed with increase in w/c, i.e., from 0.35 to 0.40. Slump values were on higher range from 25–35 mm compared to other w/c. This decrease in workability is due to the high surface area of iron ore mine tailings. The desired slump value as per the mix design for M40 grade concrete is 25–50 mm. The concrete is workable up to 40% replacement of IOT; thereafter, the mix becomes rigid due to high water absorption of IOT.

7.4.3 Compressive Strength

The compressive strength was determined using compressive testing machine with a loading capacity of 2000 kN as per IS 516-1959. The loading rate applied in the compressive was 140 kg/cm^2/min.

Compressive strength is calculated using equation 7.1.

$$f_r = \frac{P}{A} \tag{7.1}$$

where, P = load at failure; A = area of cross section

7.4.3.1 Effect of WR in Compressive Strength

At 0.35 w/c, the strength gradually increased with increased percentages of WR up to an optimum percentage of 50% (Figure 7.1). At 3, 7 and 28 curing days, maximum (optimum) strength observed with reference to the control mix concrete increased by 12.37%, 34.79% and 26.09% at WR4-40; and at 0.40 w/c, with reference to the control mix, optimum strength was observed at WR5-30. Strength

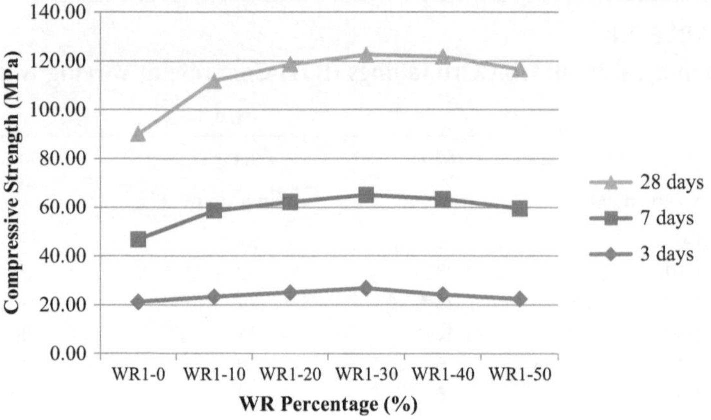

FIGURE 7.1 Compressive strength of WR concrete for w/c – 0.35.

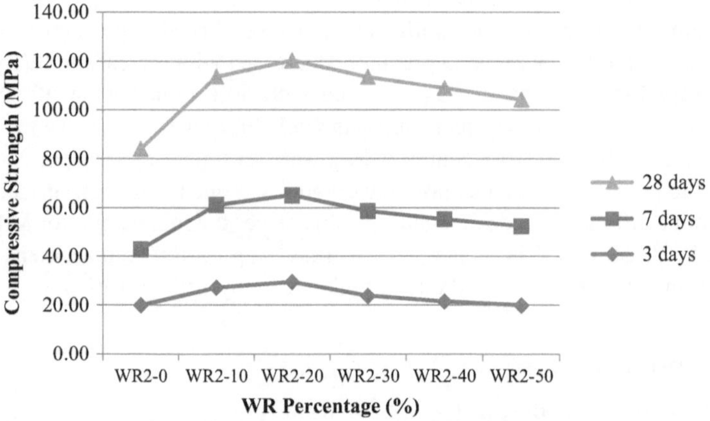

FIGURE 7.2 Compressive strength of WR concrete for w/c – 0.40.

increased by 16.33%, 34.57% and 24.79% at 3, 7 and 28 curing days as shown in Figure 7.2. Similar increase in strength was observed at 0.45 w/c also. In comparison within w/c, compressive strength decreased with increase in w/c from 0.35 to 0.45. The strength obtained with WR is greater than the control mix with reference to all the mixes. This increase in strength is due to the shape of the aggregates which fills the voids and thus binds the concrete and in turn increases the strength of the concrete (Figure 7.3).

7.4.3.2 Effect of IOT in Compressive Strength

Compressive strength of concrete was determined for varying percentages of IOT, i.e., 0%, 10%, 20%, 30%, 40% and 50% at 3, 7 and 28 curing days for varying w/c of 0.35, 0.40 and 0.45. Compressive strength shows an increasing trend in strength with

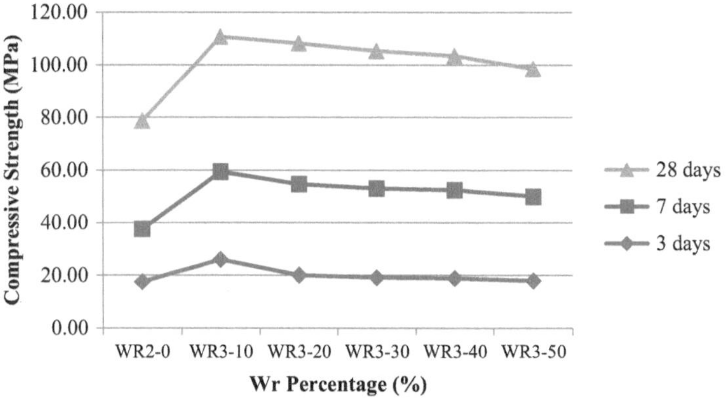

FIGURE 7.3 Compressive strength of WR concrete for w/c – 0.45.

reference to 3, 7 and 28 days cured specimens. Results at 0.35 w/c for 3 days cured specimens with reference to control mix concrete, strength increased by 15.51%, 27.00%, 37.07%, 19.87%; for 7 days cured specimens, strength increased by 10.54%, 24.52%, 31.56%, 9.46%, 6.51% and for 28 days cured specimens, strength increased by 10.88%, 14.39%, 17.14%, 8.19% and 2.02% at 10%, 20%, 30%, 40% and 50% IOT replacement, respectively. Maximum strength of 52.11 MPa was observed for 28 cured specimens and the optimum percentage obtained was at 30% IOT replacements as illustrated in Figure 7.4. Similar trends were observed for 28 days cured specimens at 0.40 and 0.45 w/c. Maximum strength of 44.85 MPa and 43.15 MPa was observed at optimum IOT percentage of 20% and 10%, respectively, as illustrated in Figures 7.5 and 7.6. In comparison within the w/c, the strength reduction was observed from 0.35 to 0.45 w/c due to the increased surface area of concrete with the increasing percentage of IOT, which in turn increases the demand of water and decreases the strength of concrete (Figures 7.4–7.6).

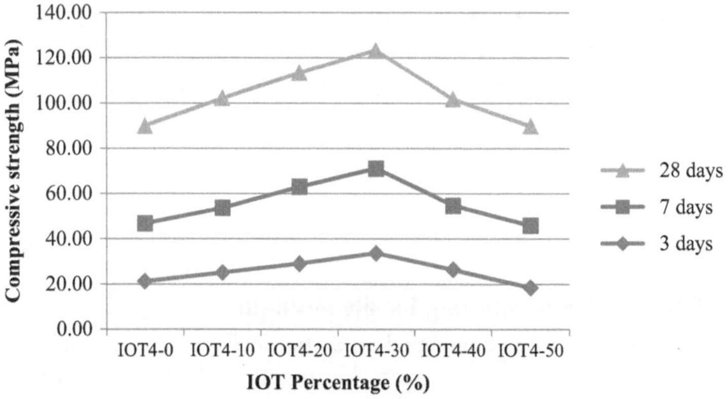

FIGURE 7.4 Compressive strength of IOT concrete for 0.35 w/c.

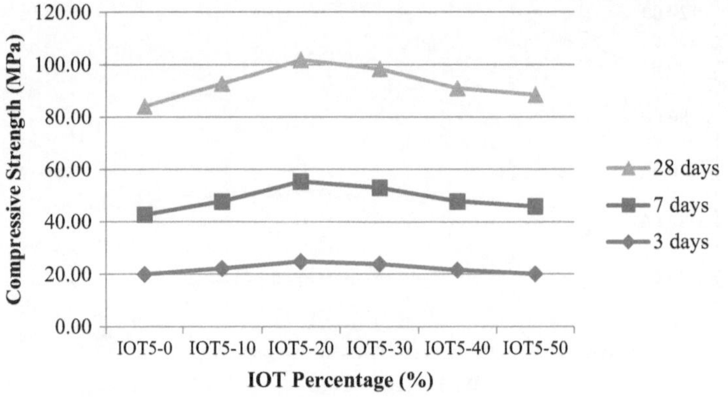

FIGURE 7.5 Compressive strength of IOT concrete for 0.45 w/c.

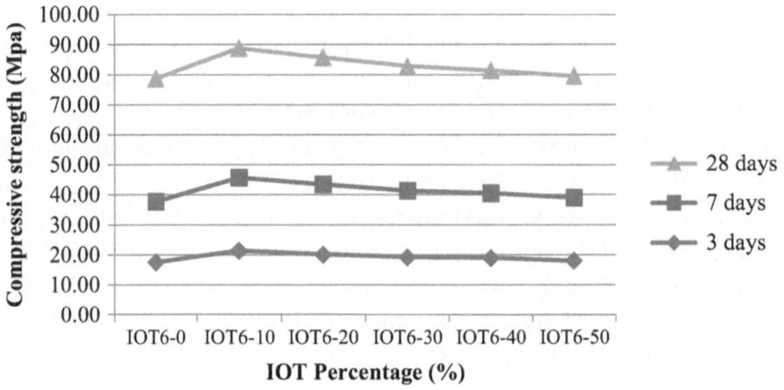

FIGURE 7.6 Compressive strength of IOT concrete for 0.40 w/c.

7.4.4 Splitting Tensile Strength

The splitting tensile strength was determined using compressive machine with a loading capacity of 2000 kN. The test is conducted as per IS: 5816-1999. The loading rates applied in the splitting tensile strength were 1.2–2.4 MPa/min.

Splitting tensile strength was calculated using equation 7.2.

$$f_r = \frac{2P}{\pi LD} \tag{7.2}$$

where, P = load at failure; L = length of cylinder; D = diameter of cylinder.

7.4.4.1 Effect of WR in Splitting Tensile Strength

Splitting tensile strength was conducted on cylindrical specimens of dimension 150 × 300 mm. The variation in splitting tensile strength with WR content is similar to that observed in case of compressive strength. The trend is shown in Figure 7.7. Concrete specimens were cured for 28 days, tested and the results are illustrated in

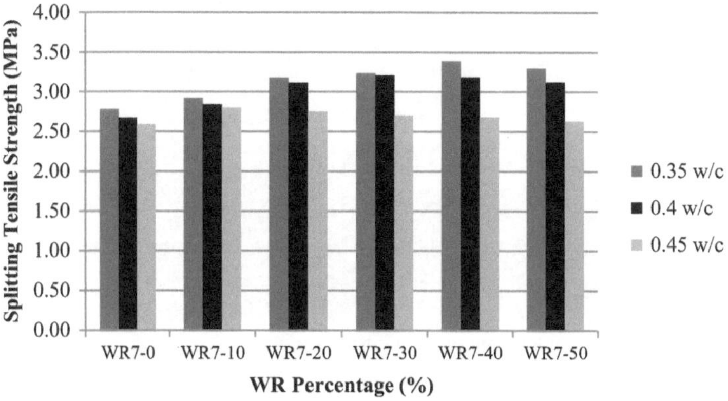

FIGURE 7.7 Splitting tensile strength of WR concrete for 0.35, 0.40 and 0.45 w/c.

Figure 7.7 for different w/c of 0.35, 0.40 and 0.45. For 0.35 w/c, strength of concrete specimens increased with reference to the control mix by 4.79%, 12.58%, 14.20%, 17.99% and 15.76 for WR replacement from 10 to 50% with 10% intervals and the optimum mix was observed at WR7-40. At 0.40 w/c, an increase in strength was observed by 5.99%, 14.15%, 16.82%, 16.04% and 14.42% with reference to the control mix for 10–50% of WR replaced concrete and the optimum mix was observed at WR7-30. Similar increasing trend was observed for 0.45 w/c and the optimum mix was observed for the mix WR7-20. In comparison of w/c, maximum strength was obtained for mixes with 0.35 w/c and least strength was obtained at 0.45 w/c. However, the strength values are higher than the target strength required for M40 grade concrete.

7.4.4.2 Effect of IOT in Splitting Tensile Strength

Splitting tensile strength was tested on 28 days cured concrete specimens of dimension 150 × 300 mm as per IS standards. Strength with reference to the control mix concrete found to be at 0%, 10%, 20%, 30%, 40% and 50% IOT replacement, respectively, were 2.78 MPa, 3.74 MPa, 3.82 MPa, 3.91 MPa, 3.86 MPa and 3.84 MPa at 0.35 w/c; 2.67 MPa, 3.62 MPa, 3.71 MPa, 3.68 MPa, 3.62 MPa and 3.58 MPa at 0.40 w/c and 2.59 MPa, 3.554 MPa, 3.51 MPa, 3.47 MPa, 3.42 MPa and 3.38 MPa at 0.45 w/c. Maximum strength was observed at 30%, 20% and 10% of IOT replacement for 0.35, 0.40 and 0.45 w/c. In comparison within w/c, splitting tensile strength gradually decreased with increase in w/c from 0.35 to 0.45. Strength obtained from control mix was in the range of 2.59–2.78 MPa and maximum strength was in the range of 3.54–3.91 MPa as shown in Figure 7.8.

7.4.5 Flexural Strength

The IOT-alccofine concrete beams were tested under three-point loading method for flexural strength. The load capacity of flexural testing machine is 230 kN. The loads

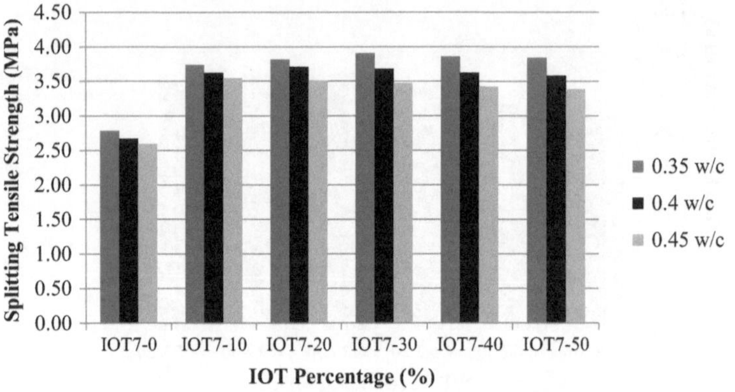

FIGURE 7.8 Splitting tensile strength of IOT concrete for 0.35, 0.40 and 0.45 w/c.

were positioned within the middle third 150 mm of the specimen, thus maintaining a loading span of 450 mm during the test. The Modulus of Rupture (MOR) of the beams was determined at 28 and 56 days cured samples after the test depending on the place of occurrence of the failure fracture.

Under third-point loading, two scenarios are possible;

a) For fracture occurring within the middle third, the MOR is calculated as:

$$f_b = \frac{\text{Pl}}{\text{bd}^2} \text{ for } a > 133.3\text{mm} \tag{7.3}$$

b) For fracture outside the middle third

$$f_b = \frac{3\text{Pa}}{\text{bd}^2} \text{ for } 110\,\text{mm} < a < 133.3\,\text{mm} \tag{7.4}$$

where, a = distance between the line of fracture and the nearer support; P = maximum load; l = span of the beam; 'b' and 'd' are the cross-sectional dimensions.

7.4.5.1 Effect of WR in Flexural Strength

The flexural strength of concrete mixes with varying percentage of WR was determined using three-point bending load as per IS standards as discussed. Flexural strength of WR concrete was tested on beams of 100 × 100 × 500 mm dimension after 28 days curing and the results are depicted in Figure 7.9. The flexural strength increased with increase in WR percentage. At 0.35 w/c for 28 cured specimens, the maximum strength of 5.32 MPa was observed at 40% replacement (WR8-40). At 0.40 and 0.45 w/c, the optimum percentage replacement observed was at 30% and 20% with 5.14 and 5.13 MPa which is higher than the control mix. The flexural strength observed is greater than the target strength required (4.5 MPa).

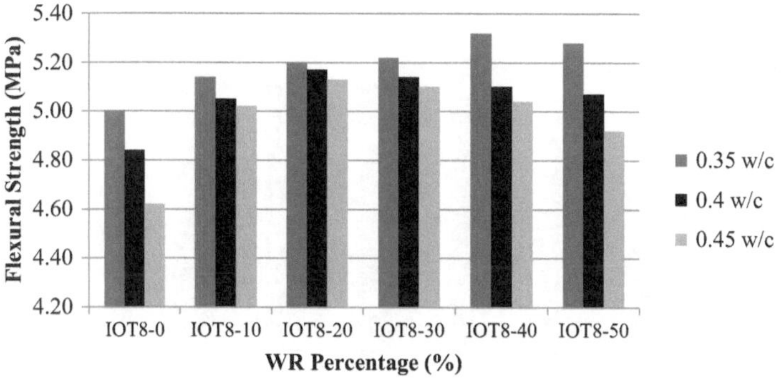

FIGURE 7.9 Flexural strength of WR concrete for 0.35, 0.40 and 0.45 w/c.

7.4.5.2 Effect of IOT in Flexural Strength

It was observed that the materials consist of major elements viz., high percentage of silicon dioxide (SiO_2), iron oxide (Fe_2O_3), aluminium oxide (Al_2O_3) and calcium oxide (CaO).

Flexural strength of all the concrete specimens with IOT was determined on concrete beams for 28 cured specimens using three-point bending load as per IS standards. The results are depicted in Figure 7.10. Flexural strength increased with increase in IOT replacement from 10%–50% with 10% interval. At 0.35 w/c, strength increased with reference to control mix concrete by 4.40%, 5.12%, 4.03%, 3.47% and 5.12%. At 0.40 w/c, with reference to the control mix, strength increased by 6.74%, 7.10%, 6.02%, 4.72% and 3.39%. Similarly, at 0.45 w/c, strength increased by 9.41%, 7.97%, 6.29%, 4.55% and 2.94% with reference to control mix. Optimum percentage was observed at 30%, 20% and 10% IOT replacement for 0.35, 0.40 and 0.45 w/c. The flexural strength observed is greater than the required target strength of 4.5 MPa.

7.5 SUMMARY

Physical properties of OPC 53 grade was used as binder material and river sand, IOT, crushed granite, WR were used as aggregates and these materials conform the IS codes and are within the limits. The specific gravity of IOT is 3.31, is high compared to river sand and has water absorption of 2.29. IOT has high surface area due to which it absorbs high water content in concrete. Chemical composition of the materials, i.e., WR and IOT were determined using X-ray fluorescence (XRF) and it was observed that the materials consist of major elements viz., high percentage of Silicon dioxide (SiO_2), iron oxide (Fe_2O_3), aluminium oxide (Al_2O_3), calcium oxide (CaO). Based on the fresh concrete slump test with reference to the control concrete mix, workability decreased with increase in WR percentage from 10% to 50% with 10% intervals. This may be due to the shape of aggregates which are angular. In case of IOT, workability decreased with increase in IOT replacement from 10% to 50% with 10% intervals due to the high surface area which absorbs more water content. Super plasticizer was added as water-reducing agent to increase the workability of

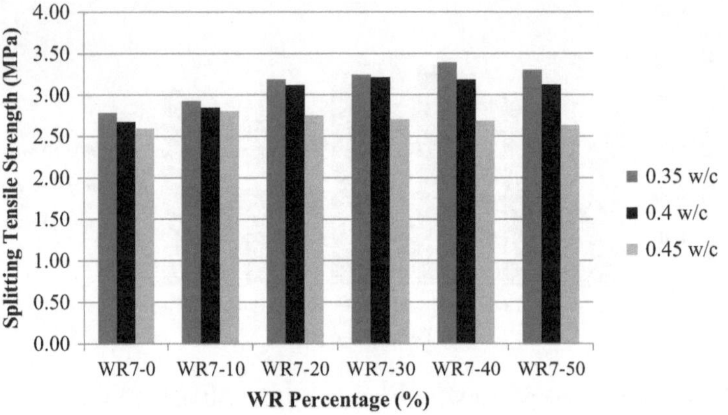

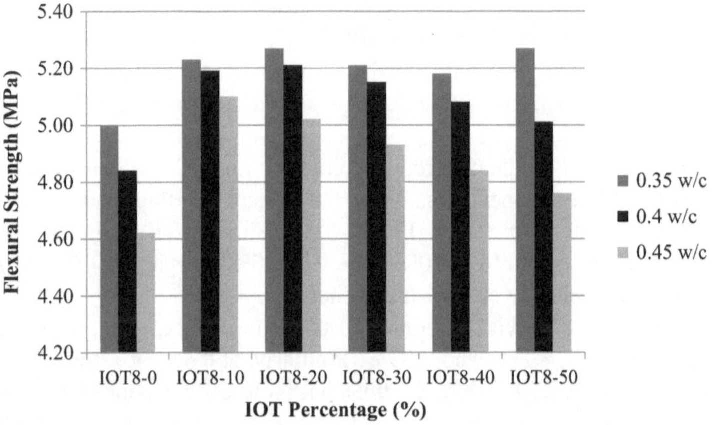

FIGURE 7.10 Flexural strength of IOT concrete for 0.35, 0.40 and 0.45 w/c.

concrete, and hence, WR and IOT concrete resulted in slump values between 25 and 50 mm. Further, based on compressive strength of WR and IOT mixes, strength increased with increase in WR quantity from 10% to 50% with 10% intervals with reference to control concrete. Strength increased by 26.09%, 24.79% and 23.29% for 40%, 30% and 20% WR replacement as coarse aggregates in concrete for 0.35, 0.40 and 0.45 w/c and for 28 days cured specimens. In case of IOT concrete, strength increased with increase in IOT content up to 30%, 20%, 10% and then the strength decreased with increase in IOT content for 0.35, 0.40 and 0.45 w/c. This is due to the high specific gravity of IOT in concrete. For 28 days cured samples, splitting tensile strength and flexural strength resulted in increasing strength up to optimum percentages as in the case of compressive strength in both the cases, i.e., WR and IOT concrete. Flexural strength obtained satisfies the mix design criteria.

From the present study, it can be concluded that iron ore WR and IOT can be used as a marginal material in construction industry. Based on the mix design for M40 grade, the strength obtained for the varying mix proportions increased gradually compared to the control mix. Hence, WR and IOT can be replaced partially as coarse and fine aggregates in concrete construction industry without compromising the strength parameters.

REFERENCES

Bureau of Indian Standards (BIS). Methods of sampling and analysis of concrete. IS: 1199: New Delhi, India; 1959.

Bureau of Indian Standards (BIS). Specification for coarse and fine aggregates from natural sources for concrete. IS: 383: New Delhi, India; 1970.

Bureau of Indian Standards (BIS). Splitting tensile strength of concrete – Method of test. IS: 5816: New Delhi, India; 1999.

Bureau of Indian Standards (BIS). Concrete Mix Proportioning-Guidelines. IS: 10262: New Delhi, India; 2009.

Castro Bastos, Lucas Augusto de, Silva, Gabriela Cordeiro, Mendes, Júlia Castro and Peixoto, Ricardo André Fiorotti. 2016. Using iron ore tailings from tailing dams as road material. *Journal of Materials of Civil Engineering*, 28(10), 04016102. 10.1061/(ASCE)MT.1943-5533.0001613.

Edraki, M.T., Baumgartl, E., Manlapig, D., Bradshaw, D.M. Franks and Moran, C.J. 2014. Designing mine tailings for better environmental, social and economic outcomes: A review of alternative approaches. *Journal of Cleaner Production*, 84, 411–420 DOI: 10.1016/j.jclepro.2014.04.079.

Filho, Joaquim Nery Santana, Silva, Sidney Nicodemos Da, Silva, Gabriela Cordeiro, Mendes, Julia Castro and Peixoto, Ricardo Andre Fiorotti. 2017. Technical and environmental feasibility of interlocking concrete pavers with iron ore tailings from tailings dams. *Journal of Materials of Civil Engineering*, 25(7), 923–931. 10.1061/(ASCE)MT.1943-5533.0001937.

Gayana, B.C. and Ram Chandar, K. 2018. Sustainable use of mine waste and tailings with suitable admixture as aggregates in concrete pavements- A review. *Journal of Advances in Concrete Construction*, 6(3), 221–243. https://doi.org/10.12989/acc.2018.6.3.221

Kang, H.Z., Jia, K.W. and Yao, L. 2011. Experimental study on properties of concrete mixed with ferrous mill tailings. *Applied Mechanics and Materials*, 148–149, 904–907. DOI: https://doi.org/10.4028/www.scientific.net/AMM.148-149.904.

Mohammad, H. Gorakhki and Bareither, Christopher A. 2017. Unconfined compressive strength of synthetic and natural mine tailings amended with fly ash and cement. *Journal of Geotechnical and Geoenvironmental Engineering* 143(7), 04017017. 10.1061/(ASCE)GT.1943-5606.0001678.

Ram Chandar, K., Gayana, B.C. and Sainath, V. 2016a. Experimental investigation for partial replacement of fine aggregates in concrete with sandstone. *Journal of Advances in Concrete Construction*, 4 (4), 243–261. DOI: https://doi.org/10.12989/acc.2016.4.4.243.

Ram Chandar, K., Raghunandan, M.E. and Manjunath, B. 2016b. Partial replacement of fine aggregates with laterite in ggbs-blended-concrete. *Journal of Advances in Concrete Construction*, 4 (3), 221–230. https://doi.org/10.12989/acc.2016.4.3.221.

Yu, L., Tian, J.S., Zhang, J.X. and Yang, R.J. 2011. Effect of iron ore tailings as fine aggregate on pore structure of mortars. *Advanced Materials Research*, 250–253, 1017–1024. https://doi.org/10.4028/www.scientific.net/AMR.250-253.1017.

Index

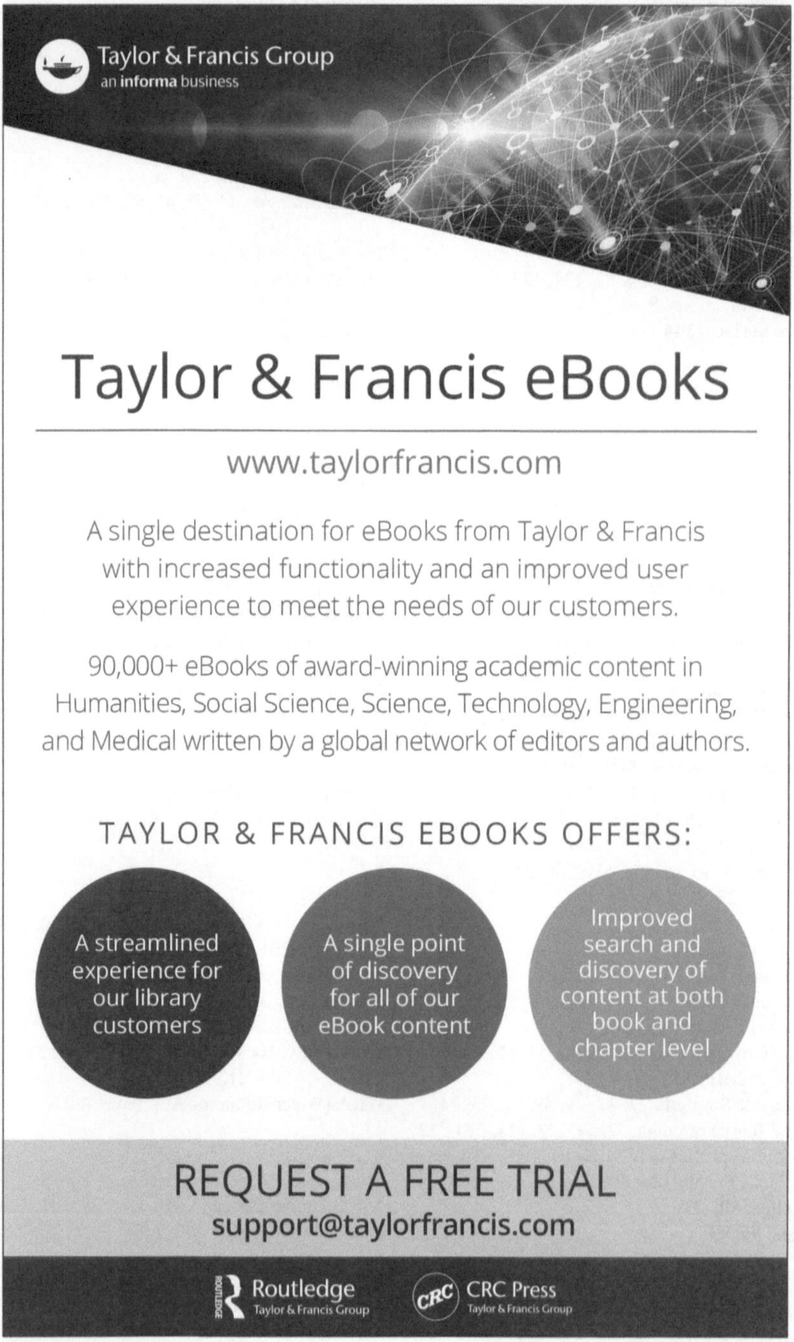
Taylor & Francis Group
an informa business
Taylor & Francis eBooks
www.taylorfrancis.com
A single destination for eBooks from Taylor & Francis with increased functionality and an improved user experience to meet the needs of our customers.
90,000+ eBooks of award-winning academic content in Humanities, Social Science, Science, Technology, Engineering, and Medical written by a global network of editors and authors.
TAYLOR & FRANCIS EBOOKS OFFERS:
A streamlined experience for our library customers
A single point of discovery for all of our eBook content
Improved search and discovery of content at both book and chapter level
REQUEST A FREE TRIAL
support@taylorfrancis.com
Routledge
Taylor & Francis Group
CRC Press
Taylor & Francis Group